"Grand Prix d'Horlogerie
de Genève"

# GRAND COMPLICATIONS

THE ORIGINAL ANNUAL OF THE WORLD'S WATCH COMPLICATIONS AND MANUFACTURERS®

## SPECIAL CHRONOGRAPHS

First published in the United States in 2006 by:
TOURBILLON INTERNATIONAL LLC
11 West 25th Street, 8th floor
New York, NY 10010
T: +1 (212) 627-7732 Fax: +1 (212) 627-9093
www.tourbillon-watches.com

CHAIRMAN Joseph Zerbib

CHIEF EXECUTIVE OFFICER & PUBLISHER Caroline Childers

EDITOR Michel Jeannot

MANAGING EDITOR Elizabeth Kindt

ART DIRECTOR Franca Vitali

DIRECTOR OF FINANCE Elliott Elbaz

In association with RIZZOLI INTERNATIONAL PUBLICATIONS INC.

300 Park Avenue South, New York, NY 10010

ISBN: 0-8478-2940-5



PRINTED IN ITALY

AP
AUDEMARS PIGUET
Le maître de l'horlogerie depuis 1875
THE WATCH OF YOUR LIFE
LE BRASSUS

"From
Woman
comes Light.
ARAGON

"Time is
the moving symbol
of a motionless Eternity.
PLATON

# CELLINI JEWELERS

New York, NY

French couturier Yves Saint Laurent once proclaimed that "fashions fade, style is eternal." After nearly 30 years, Cellini Jewelers continues to prove the truth of that insightful adage with an exquisite collection of watches and jewelry whose timeless beauty transcends fleeting trends.

Recognized as one of the world's preeminent watch dealers, Cellini's unrivaled selection of limited edition and complicated timepieces is popular among in-the-know aficionados. "We've never been interested in carrying the most watch brands. We only carry the top echelon and we stock them in-depth," says Cellini Jewelers President, Leon Adams. "We do it because it's important for our clients to see and feel the watches. Our selection also allows you to walk out of the store with the watch you want instead of placing an order and waiting weeks for it."

As more watchmakers focus on delivering exclusivity by creating small batches of limited-edition timepieces, Adams says watch collectors who face a shrinking supply turn to Cellini for dependable access to their favorite watches. "Because of our long-standing relationship and expertise in the field of high complications," Adams explains, "we are able to secure a better than average inventory of limited-edition timepieces."

Cellini's vast assortment of natural colored diamonds features an extraordinary array of pink, blue, green, orange and yellow stones.

Cellini's flagship store was established in 1977 at the Hotel Waldorf-Astoria.

Another important part of Cellini's success continues to be the company's ability to recognize and nurture young and innovative watch brands. Cellini helped introduce A. Lange & Söhne, Audemars Piguet, Breitling, F.P. Journe and more to American audiences before they were well known.

Recognizing Cellini's influence, watchmakers regularly create exclusive, special edition watches for the company. This year, Panerai will produce the Limited Edition Cellini Exclusive Radiomir featuring their new in-house mechanical movement, the P.2002, inside a rose-gold case. The special timekeeper, available exclusively at Cellini, is due out in 2007. Cellini is also teaming up with IWC for the limited edition Big Pilot Perpetual Calendar. Made only for Cellini customers, this watch combines two of IWC's specialties—the Big Pilot case and a perpetual calendar—into one exclusive watch.

## Best Of The Best

Among New York City's premier jewelers, Cellini's superb selection of rare, high-quality gems ranks as one of the best. "One of our greatest assets is the depth of our collection of diamonds and gemstones," says Adams. "It allows us to create the gorgeous simplicity of 1-carat diamond-stud earrings or something as rare as a pink diamond necklace or a canary yellow diamond ring."

To maintain the jeweler's legendary reputation for excellence, Adams personally selects all of the gemstones featured in Cellini's one-of-a-kind pieces. "I take the responsibility for maintaining our collection very seriously," he says. "Our clients expect an impeccably high standard of quality, which is why all of Cellini's diamonds are VS1 or better."

Cristina Andrews, Cellini's jewelry buyer, says she strives to create a balanced collection that satisfies tastes from understated to over-the-top. "What sets us apart is that in our display cases you will find classic white diamonds and pearls alongside a one-of-a-kind piece featuring exotic gemstones," she says. "We have an eclectic mix that appeals to everyone whether they prefer the traditional or something just off the runways of Paris or Milan."

Another Cellini specialty is its vast assortment of natural colored diamonds. Featuring an extraordinary array of pink, blue, green, orange and yellow stones, Andrews says, the quality of Cellini's diamond collection is rivaled only by museums. "This year we unveiled an amazing new Asscher-cut ring featuring two fancy colored diamonds—gray-blue and pink—surrounded by white diamond baguettes," she says. "The color and clarity of the diamonds really make this ring a knockout."

LEFT

Cellini's gemologists rigorously scrutinize each gemstone to select for maximum brilliance and create stunning, one-of-a-kind pieces.

TOP

Cellini's boutique in the Hotel Waldorf-Astoria showcases an awe-inspiring collection of fine jewels and elite timepieces.

## WATCH BRANDS CARRIED

A. Lange & Söhne
Audemars Piguet
Baume & Mercier
Blancpain
Cartier
Chopard
Chronoswiss
DeWitt
F.P. Journe
Franck Muller
Gerald Genta
Girard-Perregaux
Hublot
IWC
Jaeger-LeCoultre
Jean Dunand
Panerai
Parmigiani Fleurier
Piaget
Pierre Kunz
Richard Mille
Ulysse Nardin
Vacheron Constantin

## Personalized Style

Whether your heart's desire ticks or sparkles, Cellini's specialists provide friendly, personalized service from something as simple as watch maintenance to something as personal as a custom-designed piece of jewelry. As Cellini's gemologist, Lauren Goldsmith has helped many clients realize their radiant aspirations. "Some people come in and ask us to design something for them from start to finish. Others know exactly what they want. We handle both every day," she says. "We use the world's most beautiful gemstones to create one-of-a-kind masterpieces."

Family-owned and operated since first opening at the famed Hotel Waldorf-Astoria in 1977, Cellini's flagship remains the epitome of exceptional elegance. Its sparkling window displays of watches and jewelry have long been beautiful fixtures inside the hotel, which is a home away from home for powerful leaders and glamorous stars alike. Cellini opened a second location 10 years later with a showroom on Madison Avenue. Even in a neighborhood populated by the most exclusive luxury boutiques found anywhere, Cellini's rare treasures are a standout.

While Cellini's elite combination of watches and jewelry has drawn a crowd for almost 30 years, it is the dedicated and knowledgeable staff at both boutiques that keep people coming back.

TOP
Cellini's second boutique was established in 1987 at the epicenter of the world's most elite shopping district.

CENTER
Filled with fine jewels and a striking collection of timepieces, Cellini's Madison Avenue display windows never fail to stop passers-by in their tracks.

509 Madison Avenue at 53rd Street, New York, NY 10022
Tel: 212.888.0505

Hotel Waldorf-Astoria, 301 Park Avenue at 50th Street
Tel: 212.751.9824

800-CELLINI www.CelliniJewelers.com

CHANEL
J12
AUTOMATIC
SWISS MADE

# Letter from the Publisher

Things are getting more complex for women!

Timekeepers: what a poetic and evocative name to express the value of objects that are transformed into works of art by an elite composed of brilliant watchmakers and craftsmen! While the mechanical "keeping" of this ever-elusive time that none can hold back is already a complex task in itself, successfully revealing and sublimating it within a diminutive case is a true labor of love and artistry. If anyone needs convincing of the vocational nature of watchmaking, they need only wander through the "complications" workshops of the authentic manufactures and admire the work being done there. This experience alone is enough to help one grasp that complication watches propel watches into a dimension reaching well beyond the basic notion of an object that tells the time.

It would appear that watch manufacturers have now realized that these emotionally charged artistic objects can also stir the feminine soul. Moreover, these souls represent a clientele that is becoming increasingly fascinated by these demonstrations of mechanical expertise. The time when watchmakers imagined that only diamonds could capture women's attention and tug at their heartstrings is now ancient history!

The process initiated several years ago, whereby watch producers gradually began offering feminine mechanical timekeepers that were more than mere pint-sized versions of men's models, has suddenly picked up momentum and now offers attractive prospects both for the brands and for women themselves. While this newfound intimacy might be viewed as a case of dangerous liaisons by certain unrepentant male chauvinists, it is certainly exhilarating and thrilling enough to deserve your avid contemplation!

Caroline Childers

**CALIBER RM 011**

FELIPE MASSA FLYBACK CHRONOGRAPH

Automatic winding movement
Variable inertia rotor with ceramic ball bearings
62 jewels
Specially designed chronograph coupling system
Skeletonized titanium baseplate
Oversize date
Months at 4 o'clock with automatic adjustment for 30 or 31 days
60-minute chronograph countdown timer
12-hour chronograph totalizer
Flyback function
Water-resistant to 50 meters

Titanium caseband with bezels in either titanium, or 18-carat red or white gold

**CALIBER RM 015**
**PERINI NAVI CUP**
MARINE TOURBILLON

Manual winding movement
Carbon nanofiber baseplate
Variable inertia balance
Fast rotating barrel
Power reserve indicator
Torque indicator
Function selector
Second time zone
In line escapement
Ceramic center pivot
Winding barrel teeth and third-wheel pinion with central involute profile
Modular hand setting and winding mechanism fitted against the case back

Available in 18-carat red and white gold, or platinum

# Letter from the Editor

2007: The year of the chronograph

Several years ago, the fine watch manufactures set about tackling the major task of broadening their ranges in terms of proprietary movements. Much like engines in the automobile industry, watch calibers enable each cutting-edge "racing team" to distinguish itself from the competition by its own performances and to fine-tune its image among connoisseurs and collectors. While many of the watch engines appearing during the first years of the mechanical watchmaking revival tended to originate from the same engine specialist, things have come a long way since then.

End users are better informed and are no longer prepared to put up with a situation in which a movement powering a valuable timepiece might bear an uncanny resemblance to one found in a far lower priced segment. In order to meet these new demands and expectations, the watch manufactures have invested considerable sums of money in developing new movements, which they naturally safeguard for their own collections. This enables each company to stand out from the rest by displaying its own distinctive hallmark, not only in the aesthetics of its watches, but also through the technical approach adopted. Two, three or four barrels; exclusive escapements; original functions; varied materials: the watch industry is currently swarming with these technical "hallmarks" that contribute to a brand's image and enable it to underscore its difference from competitors. Make no mistake: this is a relatively recent phenomenon.

Over the past decade, and especially over the past five years, proprietary or manufacture-made movements have become a prominent part of Haute Horlogerie. Virtually no self-respecting manufacture can afford to be without its own caliber. Some might well say that is the least one might expect. After base movements, prestige watchmaking has entered a new and extremely stimulating stage, not only for developers but also for end customers: that of complication "in-house" movements. Within this fascinating world, the chronograph holds a place of honor. This frequently underestimated yet in fact extremely complex function is one of the most sophisticated expressions of the watchmaking art. The sheer difficulty of designing, engineering, producing and ensuring the smooth operation of a chronograph movement becomes instantly more apparent when one considers that scarcely more than a handful of companies can boast their own chronograph movement. Nonetheless, this tendency has suddenly accelerated over the past two years and a few brands have presented their very own chronographs. Moreover, the trend looks set to continue in 2007, with the announcement of several new manufacture-made chronograph movements to be presented at the spring watch fairs. Considering that the development of a single chronograph movement represents several years of work and investments amounting to four or five million Swiss francs, one gets a clearer picture of the exceptionally buoyant mood that is currently sweeping through fine watchmaking circles.

Michel Jeannot

# [ WESTIME ]

**Westime** is synonymous with quality in the realm of haute horology; from the timepieces it carries to its superior customer service, the commitment to quality is evident. This year Westime and its clients celebrate 20 years of being home to some of the finest offerings in watchmaking. Recognized globally as a premier horologer, Westime caters to everyone who shares in its passion for the art of watchmaking.

## A History of Quality

Westime, located in California, was the dream of a second-generation watchmaker. He envisioned an organization that would represent the finest in the world of watchmaking and have the client as the central focus. The company was created in support of and anticipating the renaissance of mechanical watches. There are a few featured jewelry lines, mainly those of some of the watch manufactures. However, the primary focus is on unique, one-of-a kind, rare or limited edition timepieces.

## The Extraordinary Collections

Westime offers its clients the full scope of what the watchmaking world has to offer; from traditional to technologically innovative timepieces. Some of the masters such as Audemars Piguet, Vacheron Constantin or Breguet are part of the Westime family. Revered companies such as these are the foundation for Westime's collections and the selections found here are amazing. Westime also supports contemporary visionaries. Embracing those who have captured the spirit of watchmaking, Westime supports new and independent watchmakers. For example, F.P. Journe, hD3 and Richard Mille have taken bold leaps into the future of watchmaking. These watchmakers' relentless quest to improve upon past achievements has allowed them to create some of the most breathtaking pieces in the world, such as the Journe Grande Sonnerie or the Richard Mille Architect Tourbillon.

westime
254

Offering its clients two locations in which to shop, Westime prides itself on making each customer a client by providing an unparalleled shopping experience. Westime's Los Angeles salon is a spacious and inviting haven for collectors. With over 6,600 square feet of space, the store itself is as impressive as the selections it offers. The store has a full-time watchmaker on site and handles service for most watch needs.

The Beverly Hills boutique is situated on the famous Rodeo Drive. This elegant shop caters to celebrities and tourists; it is often referred to as a "collector's candy store." Both stores have a loyal following of clients who appreciate the selection as much as the service. Haute horology sales specialists at both locations are well versed in everything from the history of the watch brands to the intricacies of complications. The staff is trained and continually updated by the various watch companies. Not only are they experts in horology, but their diversity lends to their capability. The staff hails from across the United States as well as from around the world; there are over 10 languages spoken at Westime. The service is impressive; the staff will do anything for the client. From the superior service to the elegant stores, the superb collections to the illustrious history, shopping at Westime is truly an extraordinary experience.

BRANDS CARRIED
A. Lange & Sohne
Audemars Piguet
Bell & Ross
Breguet
Breitling
Chanel
Chopard
Daniel Roth
De GRISOGONO
Eberhard
F.P. Journe
Franck Muller
Girard-Perregaux
Gérald Genta
Greubel Forsey
Harry Winston
Hautlence
HD3
Hublot
IWC
MB&F
Michel Jordi
Parmigiani Fleurier
Pierre Kunz
Richard Mille
Tag Heuer
Urwerk
Vacheron Constantin
Vertu
Zenith
10800 West Pico Blvd., #197, Los Angeles, CA 90064
Tel: 310.470.1388 Fax: 310.475.0628
254 North Rodeo Drive, Beverly Hills, CA 90210
Tel: 310.271.0000 Fax: 310.271.3091
www.westimewatches.com

CHOPARD
CHRONOMETRE

CHOPARD
CHRONOMETRE

GRAND COMPLICATIONS®

THE ORIGINAL ANNUAL OF THE WORLD'S WATCH COMPLICATIONS AND MANUFACTURERS®

TOURBILLON INTERNATIONAL, LLC
ADMINISTRATION, ADVERTISING SALES,
EDITORIAL, BOOK SALES

11 West 25th Street, 8th Floor
New York, NY 10010
T: +1 (212) 627-7732 Fax: +1 (212) 627-9093
sales@tourbillon-watches.com

CHAIRMAN
Joseph Zerbib

CHIEF EXECUTIVE OFFICER & PUBLISHER
Caroline Childers

EDITOR
Michel Jeannot

TECHNICAL WRITER
Roberta Naas

MANAGING EDITOR
Elizabeth Kindt

CONTRIBUTING EDITORS
Elise Nussbaum

ART DIRECTOR
Franca Vitali - Grafica Effe

TRANSLATIONS
Susan Jacquet

DIRECTOR OF PRESS
Maurizio Zinelli

BUSINESS INTELLIGENCE & WEB MASTER
Marcel Choukroun

DIRECTOR OF FINANCE
Elliott Elbaz

INTERNATIONAL ADVISOR
John Simonian

WEB DISTRIBUTION
www.amazon.com

PHOTOGRAPHIC ARCHIVES
Property of Tourbillon International, LLC

# Letter from the Editor

Recently, at an unveiling of new watches by one of the world's most luxurious brands, a watch engineer spoke about the difficulties and challenges involved in reducing a new movement the brand was building to 5.6mm in thickness. This particular brand had already spent more than a year in research and development of this new caliber—a column-wheel flyback chronograph with dual time-zone indication. However, the caliber wasn't thin enough for the brand's legendary status as a producer of ultra-thin movements. It took another three prototypes and an additional eight months to reduce the movement's thickness to the desired dimensions.

In this instance, many factors struck me. To begin with, this brand was releasing multiple new calibers in one year—each of which had been several years in the making and perfecting. This example of ultra-precision and determined excellence is but one of many in the world of haute horlogerie today. Increasingly, the world's finest manufactures are developing their own complicated calibers in a return to mechanical prowess that rivals the heyday of watchmaking invention.

What's more, this brand had its top engineer rather than its master watchmaker speak about the challenges that it confronted in the research and development process. In fact, more and more, as these manufactures of complicated timepieces turn to state-of-the-art technology and computers to assist in incredible fractions-of-a-millimeter calibrations, they recognize the need for top engineers to work out the behind-the-scenes details.
As timepieces become more complex, and as watch brands work with the most high-tech materials and endeavor for ultra-superior performance, the concept of building the watch is more of a team effort than ever before.

Across the board at the finest watchmaking houses in the world, engineers, designers, watchmakers and artisans of ancient crafts such as enameling and gem-setting all come together to design, develop and ultimately produce the most sophisticated complicated watches available on the market. These timepieces are master blends of micro-engineering, hand craftsmanship and studied art that bring watchmaking boldly into this next millennium and pay homage to the forefather geniuses of watchmaking. Such spectacular endeavors—fruits of a true team effort—are found in the pages of this wonderful third edition of Grand Complications.

Roberta Naas

de GRISOGONO

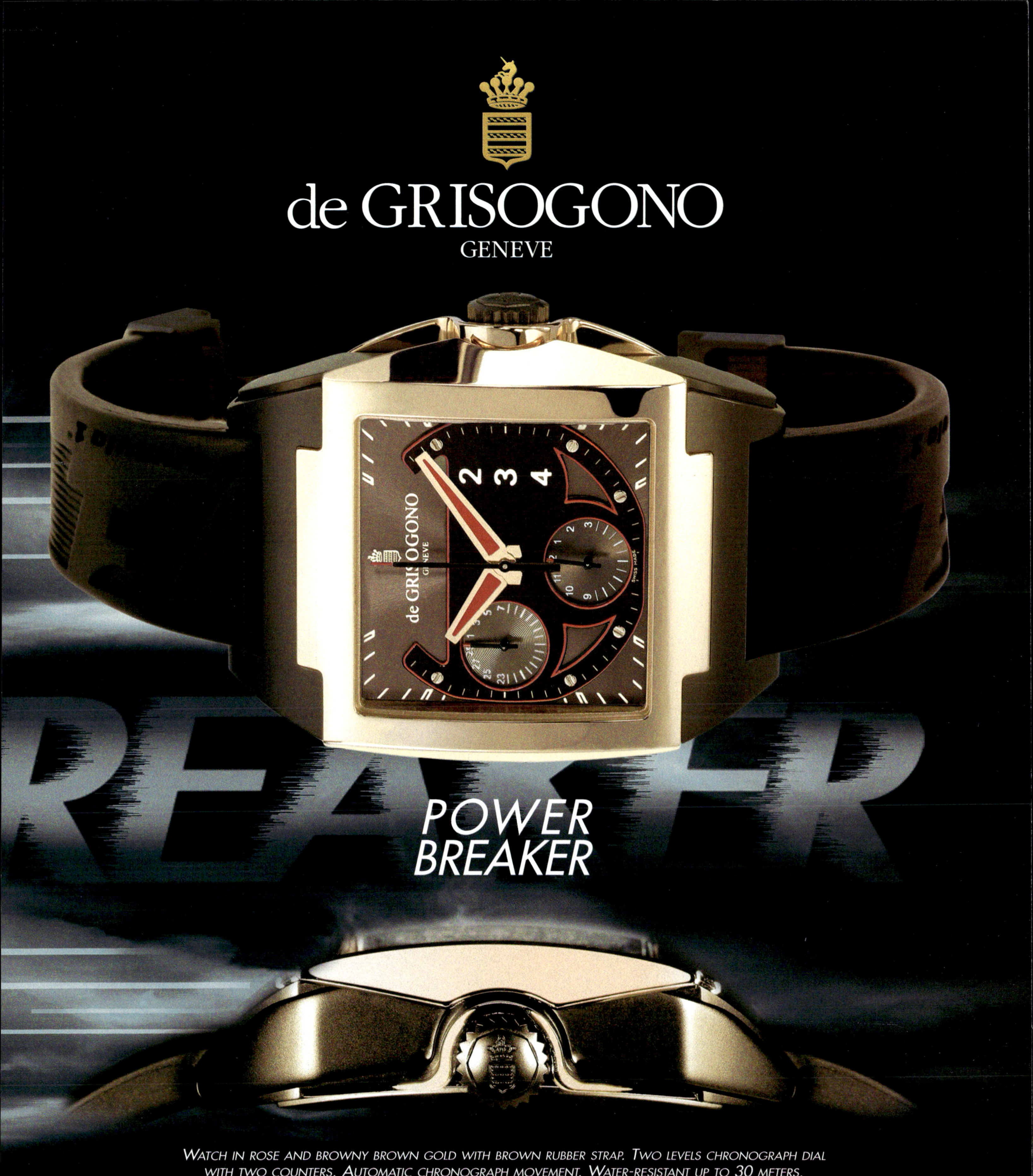

de GRISOGONO
GENEVE
POWER
BREAKER
Watch in rose and browny brown gold with brown rubber strap. Two levels chronograph dial with two counters. Automatic chronograph movement. Water-resistant up to 30 meters.

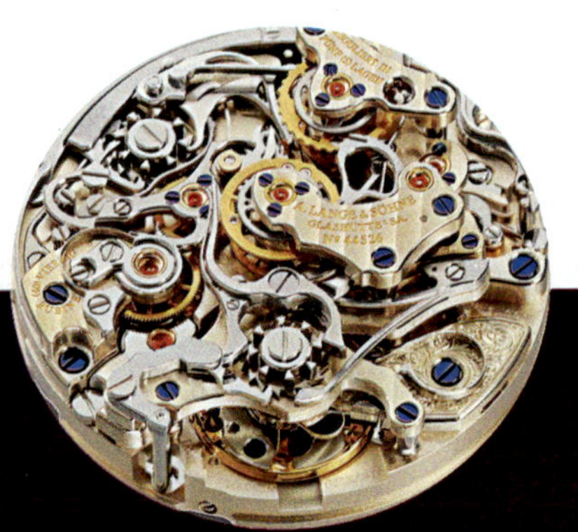

# Summary

45
15
PIAGET
30 • 20 • 10
SWISS MADE
PIAGET
BLACK TIE COLLECTION
PIAGET MANUFACTURE MOVEMENT
WORLD'S THINNEST SHAPED TOURBILLON

# Web Site Directory

| | | | |
|---|---|---|---|
| A. Lange & Söhne | www.lange-soehne.com | Maurice Lacroix | www.mauricelacroix.com |
| Audemars Piguet | www.audemarspiguet.com | MB&F | www.mbandf.com |
| Blancpain | www.blancpain.com | Michel Jordi | www.micheljordi.com |
| Bovet | www.bovet-fleurier.ch | Olivier Roux | www.olivier-roux.com |
| Breguet | www.breguet.com | Parmigiani Fleurier | www.parmigiani.ch |
| Bulgari | www.bulgari.com | Patek Philippe | www.patek-philippe.ch |
| Chanel | www.chanel.com | Piaget | www.piaget.com |
| Chopard | www.chopard.com | Pierre DeRoche | www.pierrederoche.com |
| Daniel Roth | www.danielroth.com | Richard Mille | www.richardmille.com |
| De Bethune | www.debethune.ch | TAG Heuer | www.tagheuer.com |
| de GRISOGONO | www.degrisogono.com | Ulysse Nardin | www.ulysse-nardin.com |
| Ebel | www.ebel.com | Urwerk | www.urwerk.com |
| Gérald Genta | www.geraldgenta.com | Vacheron Constantin | www.vacheron-constantin.com |
| Girard-Perregaux | www.girard-perregaux.ch | Zenith | www.zenith-watches.ch |
| Greubel Forsey | www.greubelforsey.ch | | |
| Guy Ellia | www.guyellia.com | RELATED SITES | |
| HAUTLENCE | www.hautlence.com | BaselWorld | www.baselworld.com |
| HD3 | www.hd3complication.com | SIHH | www.sihh.ch |
| Hublot | www.hublot.ch | | |
| IWC | www.iwc.ch | LVMH Group | www.lvmh.fr |
| Jacob & Co. | www.jacobandco.com | The Swatch Group | www.swatchgroup.com |
| Jaeger-LeCoultre | www.jaeger-lecoultre.com | Richemont Group | www.richemont.com |
| Jaquet Droz | www.jaquet-droz.ch | | |
| Jean Dunand | www.jeandunand.com | Auctions | www.christies.com |
| Jean-Mairet & Gillman | www.jean-mairetgillman.com | | www.sothebys.com |

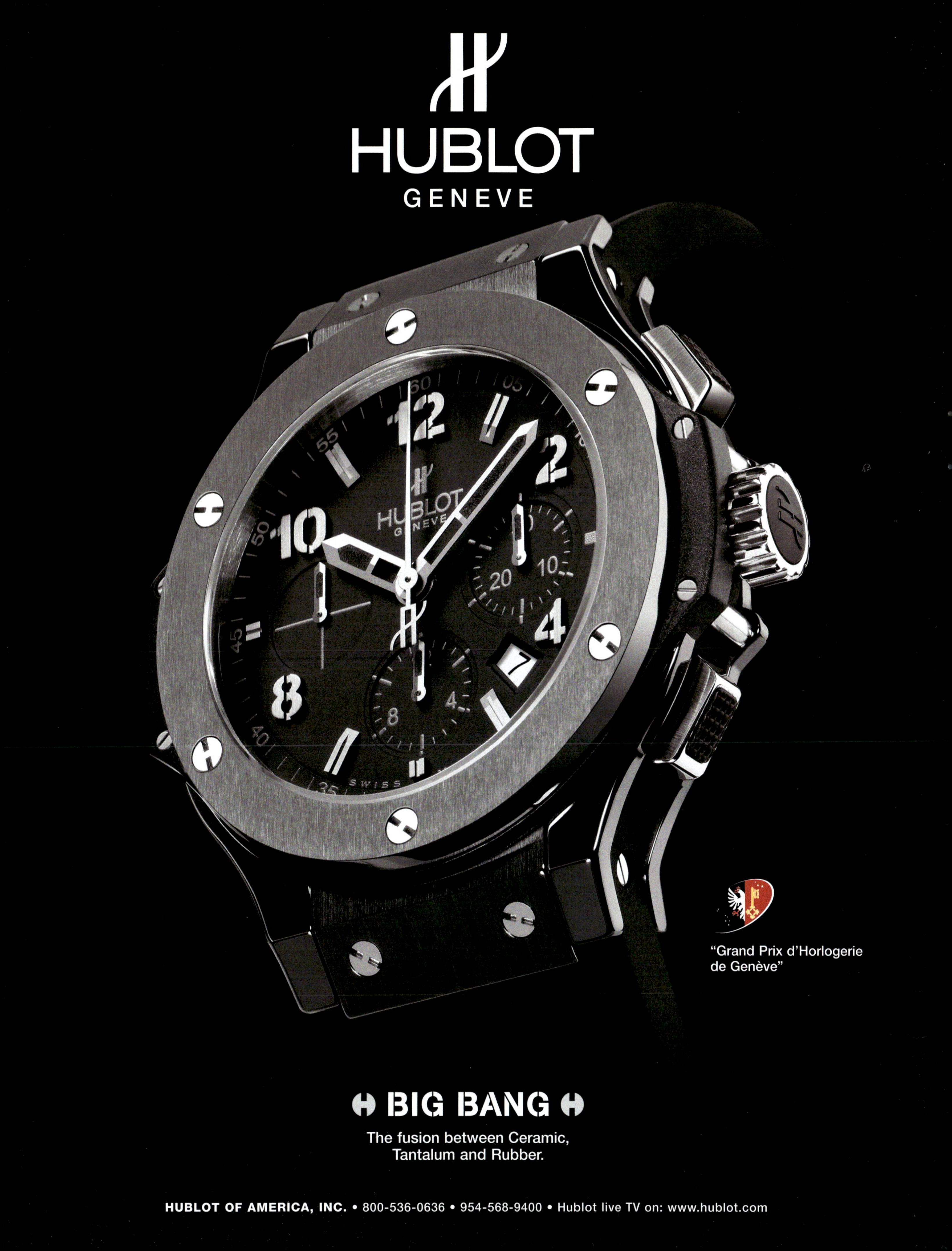
HUBLOT
GENEVE
"Grand Prix d'Horlogerie de Genève"
BIG BANG
The fusion between Ceramic, Tantalum and Rubber.
HUBLOT OF AMERICA, INC. • 800-536-0636 • 954-568-9400 • Hublot live TV on: www.hublot.com

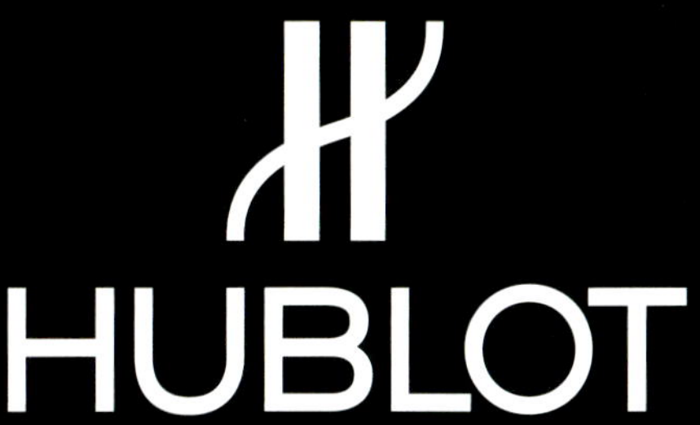

# Index

ULYSSE NARDIN
SINCE 1846
LE LOCLE - SUISSE

WHAT ARE YOU MADE OF?

# Index

AUTOMATIQUE
JAEGER-LECOULTRE
CHRONOGRAPHE
SWISS MADE

# Index

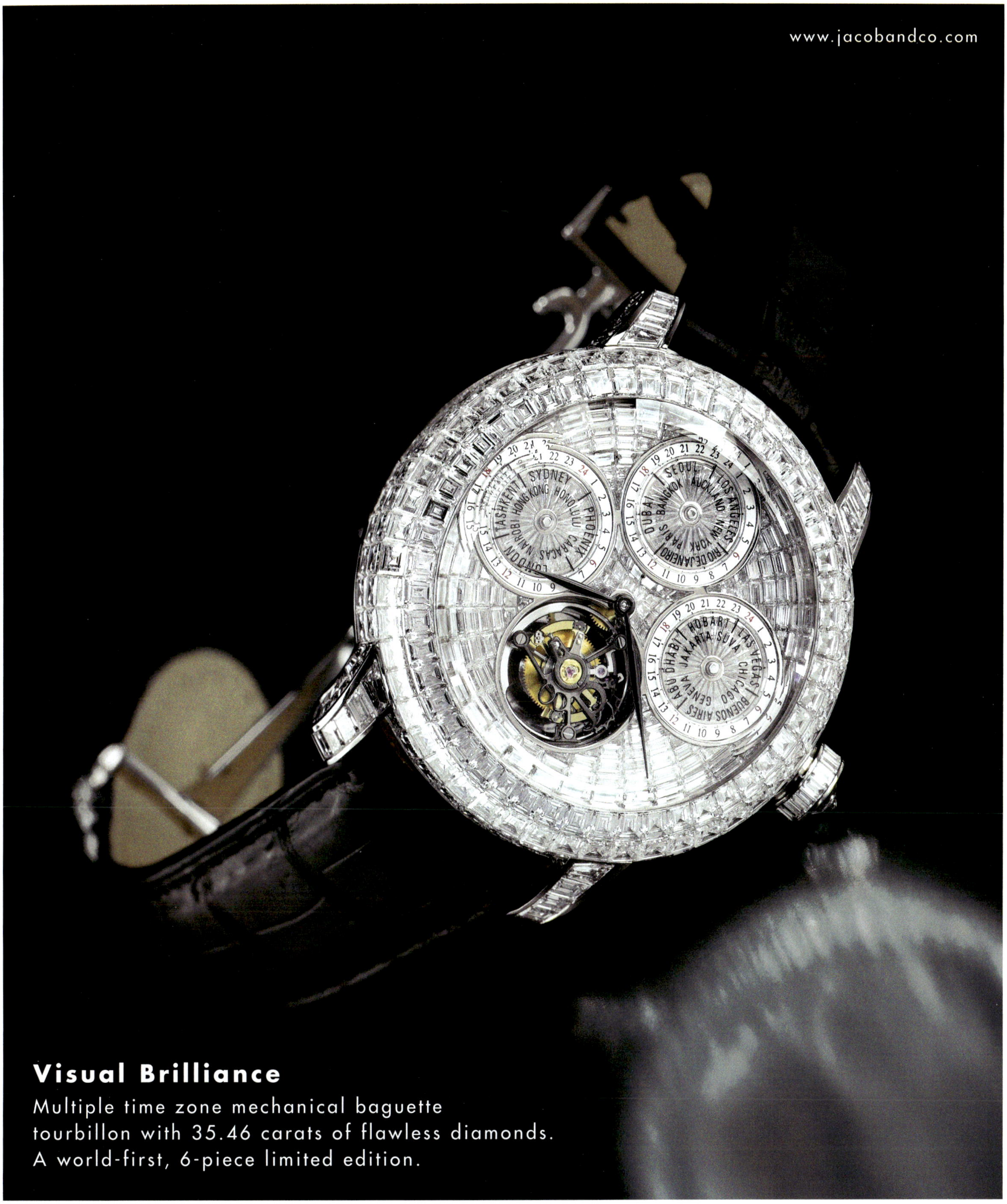
www.jacobandco.com
Visual Brilliance
Multiple time zone mechanical baguette tourbillon with 35.46 carats of flawless diamonds. A world-first, 6-piece limited edition.

IWC
SCHAFFHAUSEN
DAYS
AUTOMATIC
25
SWISS
MADE

# Engineered by India, Whiskey, Charlie.

*"Who's Charlie?"*

**Big Pilot's Watch. Ref. 5004:** Man learned to fly. And then IWC learned that man needed a special watch for the purpose. The specifications for pilot's watches still apply today: they have to be incredibly robust, antimagnetic, pressure-resistant and easy to read. IWC Pilot's Watches took their cue from the highly legible instruments in a Ju 52 cockpit.

The first Pilot's Watch from IWC appeared back in 1936. The first Big Pilot's Watch S.C., the bulkiest wristwatch ever built by IWC, came with a chunky crown and an extra-long strap that allowed pilots to wear it over their flying suits. Its precision met chronometer standards. Another legendary Pilot's Watch from Schaffhausen is the Mark 11, produced between 1948 and 1984. It was the first timepiece to feature a second, soft-iron inner case to shield the movement from magnetic fields.

**But it was even much earlier** – in 1868 to be precise – that an American watchmaker, F. A. Jones, founded the International Watch Company in Schaffhausen. Since then, IWC's engineers have developed an entire fleet of timepieces that includes the Grande Complication, the Ingenieur line, the Portuguese models and the Pilot's Watch family as well as the Da Vinci and Aquatimer series. Probus Scafusia (good, solid craftsmanship from Schaffhausen) neatly expresses the company's philosophy and, for over 100 years, has been its seal of quality. The many technical achievements and innovations that have their origins in Schaffhausen have borne impressive testimony to this for some 138 years.

*F. A. Jones.*

**The modern Big Pilot's Watch** is the sum of 70 years' experience in designing watches especially for flying. It consists of 334 individual parts and has a case diameter of 46.2 mm. The soft-iron inner case protects it against even the strongest magnetic fields. Outstanding accuracy is guaranteed by the screw balance and precision adjustment. The pawl winding system, after Albert Pellaton, is bidirectional and rapidly builds up the seven days' power reserve that can be read off on the display. Which could come in useful if it takes slightly longer than expected to obtain permission to land.

**IWC. Engineered for men.**

*IWC manufactured movement.*

*Mechanical manufactured movement | Automatic Pellaton winding system | Date display | Seven days' continuous running (figure) | Soft-iron inner case for protection against magnetic fields | Power reserve display | Antireflective sapphire glass | Water-resistant to 60 m | Stainless steel*

IWC
SCHAFFHAUSEN
SINCE 1868

IWC North America 645 Fifth Avenue, 5th Floor New York NY 10022. www.iwc.com

# Index

# The first flyback chronograph with double-rattrapante and disengagement mechanisms.

**The Lange Double Split.**
With its unprecedented complexity, the Lange Double Split is the first mechanical timekeeping instrument which can compare the duration of events that last up to 30 minutes.

The Lange Double Split is considered to be one of the most complicated watches ever crafted by "A. Lange & Söhne". Rightly so, because this rattrapante chronograph masters a feat that other wristwatches can't. Thanks to a second rattrapante hand above the chrono minute counter hand, it can stop two times in the 30-minute range rather than just in the 60-second range. While its mass is a scant 5.4 milligrams, the significance of this hand is worth far more than its weight in gold. Quite modestly, it writes a new chapter in the history of precision watchmaking. When the rattrapante button is pressed,

**Lange Uhren GmbH, D-01768 Glashütte, Germany, www.lange-soehne.com – For a complimentary catalog and DVD and for your nearest authorized**

It doesn't just sound complicated...

**The manufacture movement.** A symphony of horological artistry: the 465-part double-rattrapante calibre L001.1 movement with a balance spring developed and manufactured in-house.

the two rattrapante hands stop but the chronograph hands continue to move. Critical readers might wonder about the merits of such an instrument. The answer is simple: The Lange Double Split is the first mechanical wristwatch that provides comparative measurements of events that last as long as half an hour – to an accuracy of one-sixth of a second. Additionally, it gives watch lovers the exclusive assurance that they are wearing a timepiece which makes the seemingly impossible possible... and defines the new benchmark in precision watchmaking. As befits the Lange claim: **State-of-the-art tradition.**

dealer, please call A. Lange & Söhne, general agent: Richemont North America Inc., 645 Fifth Avenue, 10022 New York, NY, phone (212) 891 23 55.

# Index

PIERRE DEROCHE
Swiss Made
Vallée de Joux
Chronographe Concentrique
La Passion en héritage
A world première signed by
Pierre DeRoche that evolutionises
the universe of the mechanical chronograph.
This highly ingenious mechanism displays
the time measured on a single subdial
thanks to three concentric hands (hours, minutes
and seconds). A fine accomplishment
destined for connoisseurs, housed within
an extremely refined timepiece.
PIERRE DEROCHE
VALLÉE DE JOUX
PIERRE DEROCHE SA . LE REVERS 1 . 1345 LE LIEU . SWITZERLAND . T +41 21 841 11 69 . PIERREDEROCHE.COM
DARWEL/LOUP

# VERTU

## A PHILOSOPHY OF PERFECTION

**WE established Vertu to create an entirely new kind of crafted, luxury phone: a truly exceptional handset. We combined world-class expertise in mobile technology with the outstanding craftsmanship of seasoned watchmakers and jewelers. Every component in every handset is exclusively made to Vertu's own design, and put in its exact place by hand.**

***—Frank Nuovo***
***Principal Designer, Vertu***

## THE BEGINNING

Vertu was born from an obsession to create a personal communication instrument that deployed craftsmanship and technology in a way that had never been achieved before. With its global launch in 2002, Vertu pioneered an entirely new category in mobile communications with the world's first highly crafted luxury mobile phone and established itself as a true luxury brand offering exquisite products that reflect status, lifestyle and the desire for exclusivity.

## THE INSPIRATION

Vertu looked beyond the world of mobile phones and found inspiration in the finest watches, exquisite jewelry and most exhilarating cars. In doing so, Vertu took the finest materials from the worlds of watchmaking, automotive, aeronautics and jewelry and pushed manufacturing techniques to new levels. With superior materials as their foundation, Vertu handsets fuse the best in hand craftsmanship, precision engineering, tried and tested technology, high performance and personal service. The result is a timeless example of the never-ending human search for perfection.

## THE COLLECTION

Vertu handsets are the virtuoso of the mobile world and are available in three distinct collections. Within these three main collection offerings there are many Vertu variants as well as Limited Edition handsets.

## SIGNATURE

### A GRAND COMPLICATION

Vertu Signature is undeniably the finest handset ever made. Inspired by the exacting standards and tradition of grand complication timepieces, Signature is a masterpiece of craftsmanship. Signature is fashioned with precious materials such as platinum, gold, and diamonds; its face features the largest piece of complex cosmetic sapphire crystal used anywhere in the world. Each Signature handset includes 388 mechanical components with a total of 74 patents protecting its intellectual property. Signature is so complex that it takes seasoned craftspeople up to three years to learn how to assemble one from start to finish. As a result, there are only a handful of people in the world who can do it.

SIGNATURE

ASCENT

## ASCENT

### CRAFTED FOR PERFORMANCE

Inspired by the sleek lines of classic, luxury sports cars, Ascent is as tough as it is elegant. Its soft leather trim, which is the same leather that graces the inside of thoroughbred sports cars, can resist everything from petrol to lipstick, while the custom Liquidmetal® alloy chassis is so strong—with twice the hardness of stainless steel—it can withstand being run over by a race car. Numerous design details evoke the Ascent's sports car inspiration: speaker ports look like air vents, hand-stitched leather detail recalls a steering wheel, the battery-cover lock resembles a fuel-tank cap and even the screws on the battery cover were modeled after rim screws. Every component of a Vertu Ascent is crafted for performance.

## CONSTELLATION

### A NEW DEFINITION OF WORLD CLASS

Inspired by the high style of luxury travel, Constellation demonstrates perfectly how craftsmanship transcends borders. Crafted in stainless steel or gold and backed with flawless leather, Constellation has exceptional balance-to-weight ratio. Its keypad is made from highly durable ceramic, which is similar to the material used on the hull of a space shuttle. Constellation's quad-band engine ensures connection in over 180 countries around the world and its unique, Real-Time features are designed to assist the most demanding traveler.

## VERTU CONCIERGE

Vertu offers more than the highest levels of design, craftsmanship and performance in their handsets. They also provide the highest level of personal service for all of their customers. A dedicated button on every Vertu handset connects directly to Vertu Concierge, a personal service that provides expert assistance, advice and priority booking for travel, entertainment, shopping, sport, accommodation, restaurants and much more. Vertu Concierge is available around the clock and around the world, in many languages.

## THE WORLD

In 2002, Vertu Signature debuted in Paris, London, New York, Beverly Hills, Hong Kong and Singapore. Today, Vertu has its own brand boutiques and can be found in the most select outlets, including high-end department stores as well as the finest watch and jewelry stores, in 37 countries around the globe.

VERTU CONSTELLATION: SIMPLY WORLD CLASS

he new Vertu Constellation. Finished in leather, like the world's finest luggage.
rafted from stainless steel and sapphire crystal, like the world's finest watches
Twenty-four hour concierge service, like the world's most exclusive hotels.
And connected in 180 countries, like the world's most successful people.

# Index

MOD. 64 RT RA ARMOURED
CONSENTE DI RICARICARE SESSANTAQUATTRO OROLOGI DA POLSO AUTOMATICI. HA UN SISTEMA RIVOLUZIONARIO DI APERTURA TRAMITE CHIAVE OTTICA OLTRE AD UN AMPIO VANO SEGRETO AD APERTURA MECCANICA.
TO WIND UP SIXTY-FOUR AUTOMATIC WRISTWATCHES WITH A REVOLUTIONARY ELECTRONIC KEY SYSTEM AND A BIG SECRET COMPARTMENT TO CONTAIN WATCH ACCESSORIES WITH MECHANIZED SYSTEM TO BE OPENED.
SCATOLA del TEMPO®
the first, the only, the original one.
Made in Italy

THE FRENCHWAY TRAVEL: 11 WEST 25$^{TH}$

TEL: 1.212.243.3500 • FAX: 212.243. 3535 • TOLL-FREE

"Kiss Kiss darling,
where do you want to go?"
izak

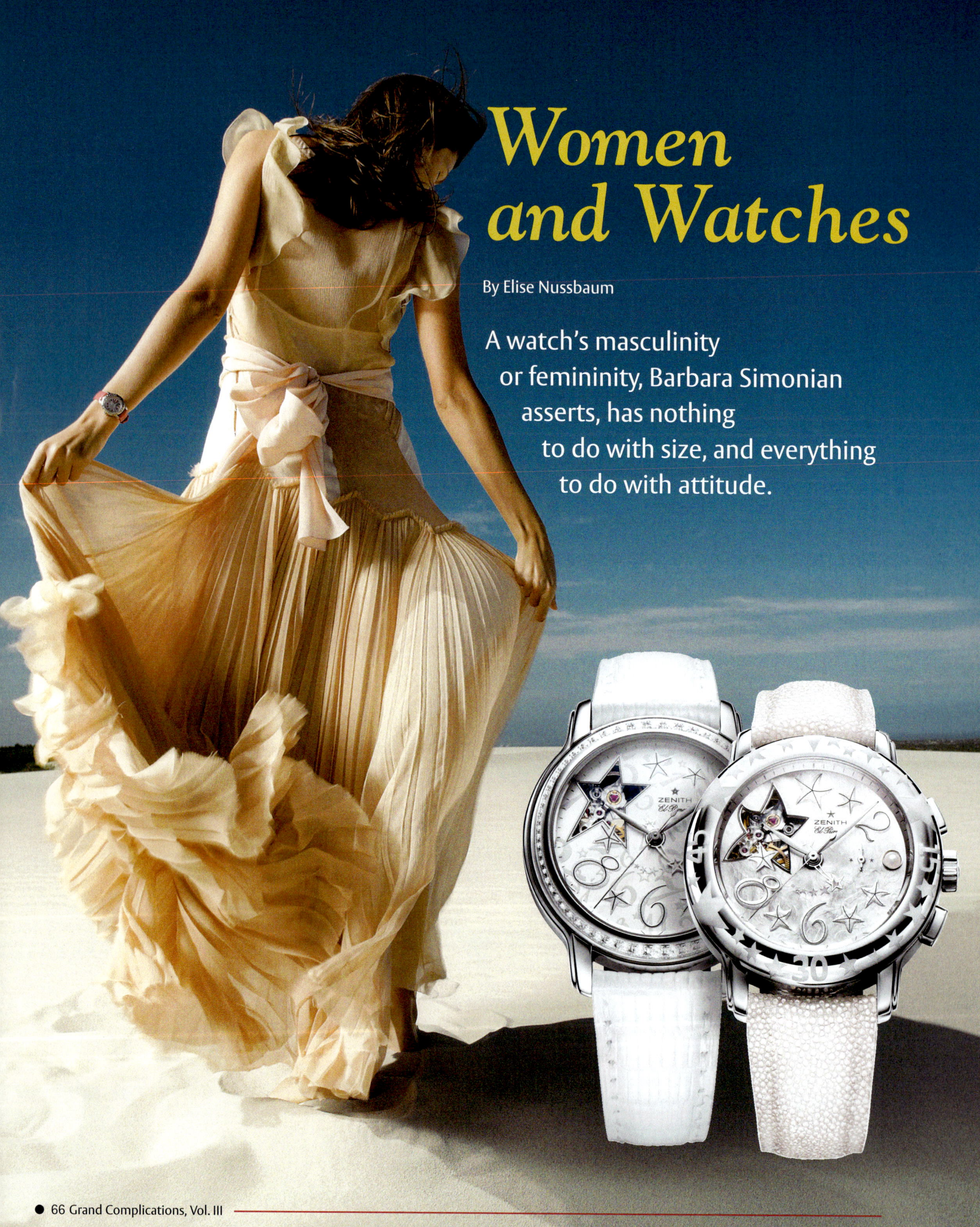

# Women and Watches

By Elise Nussbaum

A watch's masculinity or femininity, Barbara Simonian asserts, has nothing to do with size, and everything to do with attitude.

## —It's not just about diamonds anymore.—

Barbara Simonian, co-owner of Westime Watches, talks about women, watches, and why we won't take "pink" for an answer.

"A watch is like clothing," she explains. "You have to feel comfortable with it. I wear a big watch and it looks good on me. A lot of women wear men's watches—they look good and they feel comfortable." This forward-thinking attitude is just one sign of a sea change taking place among the relatively small—but rapidly growing—group of female watch aficionados. Now more then ever, as women reap the benefits of thirty years of increasing opportunity in the work force, they are asserting themselves across the board, a trend mirrored (and perhaps stimulated) by advertising campaigns exhorting women to treat themselves, such as DeBeers' call for women to raise their right hands (and adorn it with a diamond ring).

New complicated watches aimed at women are picking up on this trend, but as Simonian remarks, we still have a ways to go. The mistake of many watchmakers, she reveals, is that "they still think it has to be small, it has to be pink. There were watches that came out a few years in pink, pretty pastel colors—they're cute, but that's it. They're not serious. But businesswomen don't wear watches with pink dials." Watches, even the pretty ones, are moving from the province of fashionable (read: pink) accessories to sophisticated pieces of machinery that just happen to perfectly complete outfits. "Diamonds are nice," Simonian says. "I like diamonds—but it's not all about the diamonds anymore." More and more up-and-coming watchmakers are cottoning on to this new attitude among women, and it is easier than ever to find collections in which every watch, whether designed for a man or a woman, can be worn comfortably and stylishly by a woman.

Though more and more women are getting serious about watches, the high end of the market is still dominated by men. At Westime Watches, a prestigious retail boutique, the high-end complications, those that cost $100,000 and above, are still being bought almost exclusively by men. Despite the strides that have been made, Simonian says, "Women, at this point, still don't understand—'Why

should I spend that much money on a watch?' they think." Simonian is confident, however, that times are changing, and that women will arrive at those most rarified straits. In the meantime, the pieces that hold the most appeal for women are chronographs, watches with power reserve, and automatic watches. The chronograph is by far the most popular complication among women, Simonian notes, "because we understand how it works." Unlike the dazzling yet sometimes opaque workings of other complications, both the purpose and apparatus of the chronograph are fairly straightforward. Some complicated watches have skeletonized dials, which reveal the heart of the watch and the many pieces that make it, well, tick. This approach blends a sophisticated appreciation of horological technology with a visual that is increasingly chic. "In a way, it's still about the look," Simonian explains. "But it's also about being different, about expressing one's individuality." The exposed innards of a chronograph or tourbillon provide a counterpoint to any outfit that is at least as dramatic as a diamond-set bezel, and often much more original.

The modern businesswoman is the main customer for these new, feminine complications. No shrinking violet, she is active, successful, discerning—and doesn't mind displaying these qualities on her wrist. "There's still a lot of progress to be made," says Simonian. "It's still a man's world. But now there's a bit of showing off [on the part of women], like, 'Hey look, I have something that I appreciate, too.'" Like an Italian suit, a French vintage wine, or a German luxury sedan, a fine Swiss watch reveals an appreciation for the finer things in life, as well as an acknowledgment of the talent, ambition, hard work and savvy necessary to acquire them. "A watch shows the personality of its wearer," Simonian states. "The woman who buys a watch like this expects it to be noticed, expects men to ask her about it, and today we can explain, say, 'I bought it myself.' It's a good feeling." But think not that this confluence of connoisseurship, acumen and pride in one's success is an exclusively masculine domain. "I would never say," Simonian clarifies, "'We want to be like men,' or, 'We want to be the same as men.' A woman is a woman and a man is a man. This is about a new way of being a woman." And it is. It's one that eschews traditional ideas of feminine weakness, naïveté and timidity.

Women are also becoming serious collectors at a rapid pace. From her vantage point as co-owner of Westime Watches, Simonian sees more and more women buying winding boxes, many of them with more than one gear. Some of her female clients have up to ten watches—and they don't think of them as just fashion pieces. In Simonian's view, this focus on watches as

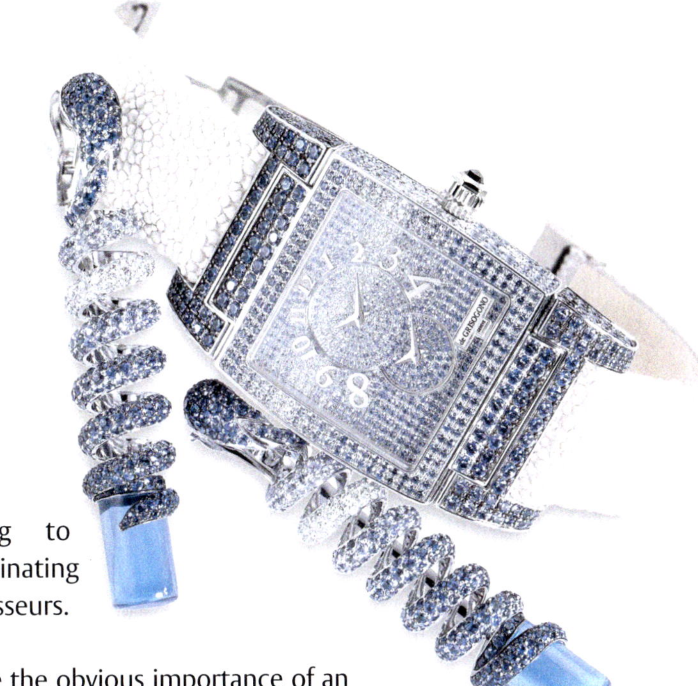

collectible works of art, as opposed to mere accessories, is reflected by a move toward increased individuality in watch choice: "We're moving away from wanting the pieces our friends have, we're focusing more on what matches us as individuals."

Faced with this surge of female demand, the watch market is undergoing some major changes of its own, albeit a bit more slowly; watches are, as Simonian points out, still overwhelmingly designed by men. "I compare it to a buying a car," Simonian says. "Twenty years ago, when I got married, a lot of women got married very young, and the husband would ask, 'What car do you want?' and the wife would say, 'Red.' Today we know which brand, which features, which tires we want." That is no accident, but the result of a clientele that has grown to like the feeling of taking matters into its own well-manicured hands. "Today," Simonian relates, "when I show a complicated watch to a women, it's not about how it looks. They want to know how it works; 'Explain it to me,' they say. The comments we used to get about mechanical watches were, 'It's too busy,' but now women are more likely to say, 'Is that handmade?'" Simonian has also done her part to educate the watch-buying public: "I love talking to people about the watches, because when I wear something, I want to be able to explain it. Understanding something and being able to explain it lead to a greater appreciation of the object." An educated consumer who knows exactly what she wants may well be the strongest impetus for change in a market that prides itself on catering to discriminating connoisseurs.

Despite the obvious importance of an educated clientele, there still exists a lack of resources available for female watch collectors. Simonian recalls an argument with a magazine publisher who was stuck in a time warp: "I said, 'We want luxury, we want to know about cars, we want to know the news about other things besides spas, or lipstick, or how to cook a roast!'" Sexist stereotypes still persist; as Simonian says, "Right now the idea is still out there that women are not interested in mechanics or complications, so there's nothing for women in that way." What is missing, it seems, is a community of woman watch collectors who could share information among themselves about exciting new products or complications. The current system of relating information, though effective and personalized, is less advertising than ad hoc. "The way it works," Simonian describes, "is a woman comes into our store looking for a watch, and we sit down with her and explain to her some of the history behind the pieces we have." Educating a customer base one client at a time may seem a painfully slow way of going about the process, but Simonian is optimistic. "Twenty years ago," she points out, "nobody knew anything about watches. We didn't have any of the watch-centered magazines and other publications that we have today. Now we have to go through that process of education again, focusing on women."

# Chronographs, Chronograph Rattrapantes and Flyback

RICHARD MILLE
TACHYMETER
RM008
HANDS
SWISS MADE

# Vice-President of Breitling

# Jean-Paul Girardin

**All too many people are doubtless not yet aware that Breitling, a Grenchen-based company with a second production site in La Chaux-de-Fonds, is the only major watch brand to achieve 100% chronometer certification by the COSC of its entire production.**

"The Navitimer represents a concentrated blend of the very essence of Breitling."

The firm that has built its reputation on four key pillars – performance, sturdiness, reliability and functionality – has also played a leading role in the field of the chronographs. It does indeed produce instruments for professionals, but these are also distinctive and remarkably reliable timepieces that meet the expectations of the most demanding connoisseurs. The growing importance of the independent brand, currently headed by the discreet company owner Théodore Schneider, is based on an extremely consistent strategy.

For Breitling vice-president Jean-Paul Girardin, "the chronograph stands out in the area of complications as a truly useful function with an immediately perceptible practical appearance. It is a real plus, in terms of both functionality and product image. The powerful aesthetics of the chronograph provide scope for countless variations linked to its functions. Chronographs are synonymous with short time measurements, competitions, surpassing oneself and records: a whole array of factors relating to the world of sports and aeronautics. These models carry undeniably dynamic connotations and this function is a real added bonus for the user".

By cultivating a longstanding and close involvement in the history and evolution of the chronograph, Breitling has made several decisive contributions to its progress. These include the first

FACING PAGE
*One of the first advertisements proclaiming the virtues of the Navitimer.*

THIS PAGE
*Breitling was a longtime supplier of high-quality cockpit chronographs.*

wristworn chronograph in 1915 and the first independent pushpiece in 1923. Breitling was also the first to give the chronograph its current configuration by creating the second reset pushpiece in 1934, and subsequently introduced the first self-winding chronograph movement for wristwatches in 1969. In a nutshell, Breitling has left its indelible imprint on the history of short time measurement.

The Breitling saga is also inextricably entwined with that of aviation. Rather than merely going with the trend flow, the company has committed itself to this sector in a lasting and concrete manner, ever since it began supplying its professional instruments to armed forces and airlines in the 1930s. The 1979 acquisition of the company by Ernest Schneider, an electronic engineer and a seasoned pilot, further reinforced these aeronautical ties. Ever since, the brand has been a partner of choice for the major international events, supports several flight formations featuring legendary aircraft or aerobatics teams, and has been associated with some truly remarkable exploits. Many will naturally recall the adventure of the Breitling Orbiter 3 in 1999, which completed the first ever non-stop round-the-world hot-air balloon flight. This feat, a new milestone in aviation history, instantly enhanced international awareness of the brand.

## Cult object

That same year, Breitling accomplished a second crucial feat in the watchmaking world: the firm chose to opt for 100% chronometer production. Breitling thus submits all its movements – both mechanical and quartz ones – to the rigorous trials imposed by the Swiss Official Chronometer Testing Institute (COSC). In order to grasp the extent of the effort required to achieve this high level of quality, one need only remember that Breitling is the only major watch brand today that has its entire production chronometer-certified. In addition to the modernization and extension of its headquarters in Grenchen, this quest for perfection, reliability and precision also called for the construction of ultra-modern building in La Chaux-de-Fonds, a unit specializing in the development and production of mechanical chronograph movements.

One of Breitling's most emblematic products is undoubtedly the famous Navitimer. First launched in 1952, this chronograph is distinguished by its circular slide rule that may be

*FACING PAGE*
*Top left*
*Since its 2004 makeover, the Chronomat is enjoying a new lease on life with the Chronomat Evolution.*

*Top right*
*The first Chronomat as produced for the Frecce Tricolori, the elite Italian aAir fForce aerobatics team in 1984.*

*THIS PAGE*
*Navitimer World: a dual-time-zone display for one of the latest models in the collection.*

used to perform all airborne navigation calculations. Rapidly adopted by many pilots around the word, the Navitimer soon became a cult object. Moreover, the fact that, after 55 years of continuous production, the Navitimer is the longest-lasting chronograph currently on the market also testifies to its success. Representing the latest newcomer to the collection, the Navitimer World displays a second time zone on a 24-hour dial by means of a highly visible hand. With its characteristic slide rule and its case measuring 46mm in diameter, the Navitimer World is the most imposing within the collection.

For Jean-Paul Girardin, Vice-President of Breitling, "the Navitimer is a concentrated blend of the very essence of Breitling. First of all, it is a mechanical chronograph; secondly, its highly innovative slide rule perfectly embodies our link with aviation. Moreover, with its 55 years of experience and its enduringly successful design, the Navitimer also represents Breitling's long-term vision and the brand's spirit of consistency and continuity. In addition, this model has a powerful design that makes it immediately recognizable even at a distance. Finally, the Navitimer is the perfect illustration of the Breitling signature commitment to constantly offering the best possible compromise between technology, functionality, precision and aesthetic appearance."

### Boldness and consistency

In additions to such consistency, Breitling also displays the kind of boldness that enables it to fly into a crosswind if necessary. Thus in 1984, when the firm launched its Chronomat model, it was both anticipating and stimulating the renewal of the mechanical chronograph. Initially adopted by the Italian Frecce Tricolori flight team, this forceful watch that is easily identifiable by the notched bezel – enabling easy handling even with flying gloves – quickly earned global success. Since 2004, this model is enjoying a new lease on life with the launch of the Chronomat Evolution. This revamped and sophisticated chronograph remains loyal to its primary vocation as a sturdy, accurate and functional instrument for professionals. "And is still the brand's best-seller in terms of volume", points out Jean-Paul Girardin.

Further reinforcing the prestigious nature of Breitling, the partnership with the distinguished English car manufacturer Bentley has given rise to the "Breitling for Bentley" collection which is attracting a good deal of attention. It represents a welcome complement to Breitling's high-end range.

*Karl-Friedrich Scheufele, CEO, Chopard.*

*FACING PAGE Some 16,000 hours of research and development went into creating the L.U.C Chrono One.*

## Co-President of Chopard

# Karl-Friedrich Scheufele

## L.U.C. Chrono One

Chopard co-president Karl-Friedrich Scheufele chose to celebrate the 10th anniversary of the Chopard Manufacture in Fleurier in style. To mark the occasion in autumn 2006, he opted to focus most of the attention on products by presenting specialists with an original and highly unconventional chronograph.

"The chronograph is an entirely underestimated complication."

The L.U.C Chrono One houses the L.U.C Caliber 10 CF, a self-winding chronograph with central rotor, for which three patents have been filed.

"The chronograph is an entirely underestimated complication," according to Scheufele. "To tell the truth, I was the first to be surprised by the truth of this fact and I underestimated the complexity and the development we would need when we launched the L.U.C Chrono One project. It is indeed no coincidence that there are still so few manufacture-made chronograph movements on the market. The number of companies currently mastering this extremely complex production is far smaller than those making tourbillons, which means that the complexity of the chronograph mechanism is a reality that bears no comparison."

This strong statement confirms that the gift Chopard Manufacture has given to connoisseurs in celebration of its 10th anniversary—the L.U.C Chrono One chronograph driven by the new L.U.C 10 CF movement—is an event that vividly illustrates just how far Chopard Manufacture has come since its creation in 1996. "This was undoubtedly the most time-consuming and most complicated project we have undertaken in the past ten years," said Scheufele. A team of 25 full-time employees in the two Research & Development departments in Fleurier and

*Top • Three patents have been filed to protect the proprietary "L.U.C 10 CF" movement equipping the new Chopard chronograph.*

*Above • The first version of the L.U.C Chrono One, in white gold, is a in is a "concept watch" edition produced in a limited series of 100 pieces.*

*FACING PAGE*
*Top left • L.U.C Chrono One, viewed from the back.*

*Top right • The L.U.C Chrono One is equipped with a self-winding column-wheel chronograph movement featuring a generous power reserve and a flyback function.*

Geneva devoted a total of approximately 16,000 hours to this project.

So how can one tell an Haute Horlogerie chronograph from a standard chronograph? Chopard's co-president has a few ideas on the topic:

"The chronograph must be integrated rather than modular, that's the first difference. It is somewhat like deciding to build a sports car: if you start from the chassis of a limousine, you are bound to end up with an unsatisfactory compromise. To build a genuine sports car, you have to start from scratch, and the same is true of a chronograph. Moreover, integrated construction is more beautiful, purer and more efficient. Another fundamental element is the column wheel. It is the nerve center of the chronograph and plays a key role in the sheer technology of the movement as well as its classical appearance. Finally, an Haute Horlogerie chronograph must stand out in terms of the inno-

vation it embodies. In this respect, the L.U.C Chrono One is particularly representative of an Haute Horlogerie model."

It is thus obvious that this timepiece, which clearly meets the above-mentioned Haute Horlogerie criteria, belongs to a proud lineage of high-end column-wheel mechanisms with flyback function, automatic winding and a substantial power reserve. Created and developed in record time, the L.U.C Chrono One has given rise to three patent requests.

The reset function is operated by a control pivoting three supple hammers, thereby enabling an optimal return of the three counter hands, and also offers the advantage of absorbing the impact of the resetting operation (patent pending). The L.U.C Caliber 10 CF also incorporates a "small seconds reset" function combined with a stop mechanism on the movement (patent pending). Meanwhile, the self-winding mechanism enables winding in both directions and comprises a unidirectional set of gears (patent pending), a system that avoids losses in energy and ensures optimal winding speed.

The integrated L.U.C Chrono One is a double-pusher column-wheel chronograph. The pusher at 2:00 controls the start and stop functions, while the one at 4:00 handles the chronograph reset, flyback and small seconds reset functions. The "stop-start" function works with clamps in conjunction with a vertical coupling-clutch, thereby making it possible to operate the start and stop functions without provoking any variation in amplitude and thus maintaining the regulating performances of the sprung balance.

In unveiling this new caliber to mark its 10th anniversary, Chopard Manufacture launched the L.U.C Chrono One "Concept Watch" in a white-gold limited numbered edition of 100 pieces. As far as the movement is concerned, these 100 watches are equipped with a pre-series "Nr. 0" movement specially handcrafted prior to industrial production—which explains why connoisseurs unable to obtain one of the first 100 are eagerly awaiting the series-production run, which should be presented at BaselWorld 2008.

## CEO of Jaeger-LeCoultre

# Jérôme Lambert

Historically speaking, the production of watch movements is one of the key points of the manufacture Jaeger-LeCoultre. In 1891, the venerable firm began filing patents covering inventions related to chronographs, and for decades it was involved in producing excellent "motors" incorporating this much sought-after complication within valuable models capable of measuring short times.

"The chronograph is one of the toughest complications to master."

During the heyday of pocket-watches and right through to the 1930s, Jaeger-LeCoultre produced some of the most superb ultra-thin calibers of the era including this function. Nonetheless, when the wristwatch took center-stage, the company in Le Sentier relied on the talents of other manufacturers to equip its own models. However, as Jaeger-LeCoultre CEO Jérôme Lambert puts it, "within the framework of the brand evolution, the mastery of complications is an ongoing daily challenge. And within this quest calling for a successful blend of innovation and technical accomplishments, the chronograph is one of the hardest objectives to reach." Thus, it was essential to resume in-house research in order to achieve complete independence in terms of mechanical production.

*FACING PAGE*
*Jérôme Lambert, CEO, Jaeger-LeCoultre.*

*THIS PAGE*
*The Master Compressor Chronograph is distinguished by its famous highly water-resistant crown system.*

By the time Jaeger-LeCoultre presented the rectangular shaped hand-wound chronograph caliber for the Reverso, Jaeger-LeCoultre could legitimately claim to master the whole spectrum of complications. Despite this, the range still lacked a mechanical component intended for watches in the Master 1000 Hours collection, which had for a long time merely made do with a version by a

*THIS PAGE*
*Top left • To launch a timing operation with the AMVOX2, the wearer need only press the crystal at 6:00.*

*Top right • Back of the Reverso Squadra World Chronograph,*

*Above • With the AMVOX2, the brand has pushed the style envelope a little further.*

*FACING PAGE*
*Top center • The Reverso Squadra Chronograph was first presented to the public in 2006.*

*Top right • The Master Compressor Extreme World Chronograph with black dial.*

high-quality mechaquartz movement. We are reminded by Lambert, who is also an informed connoisseur of watch production, that "creating a chronograph implies the ability to master significant constraints in terms of weight, pressure, resistance and operational equilibrium. Only a long learning and experience curve, as well as expertise in each of the sectors that contribute to producing a watch, can enable a company to master all the constraints involved in the series production of a chronograph caliber." In saying, this, Lambert is quick to add that "the small number of manufacture-made models on the market is sufficient proof of the complexity of such developments".

### Broad scope for development

After many long years of testing, the new in-house mechanical self-winding chronograph was finally presented in 2005. The Extreme World Chronograph took its place within the Master Compressor collection and is thus expressed in a case carefully designed to cope with the most stringent constraints. In this respect, the instrument fully matches the convictions of the CEO who recently reiterated his conviction that "the chronograph still has tremendous scope for development and applications." Endowed with a calculating function conferring technical instrument status, this watch naturally deserved a case on a par with its mechanical capacities. "Here at Jaeger-LeCoultre, we intend to explore new uncharted territories for this timepiece," said Lambert recently. Upon hearing this bold assertion, one infers that he is speaking of whole new fields of adventure, especially when one considers the extent to which the Extreme World Chronograph has been "built" to meet most situations encountered by active men.

But the youthful CEO also had other things in mind, and clearly intended to speak about technical refinements designed to "clothe" this movement when he announced that, with the AMVOX 2, the manufacture has taken the stylistic exercise a step further. Thus, on this innovative product, the pushbuttons symbolizing the very identity of the product are simply non-existent. Within this new configuration, all the wearer needs to do is press the glass at 12:00 to start a timing operation; press in the same place to stop it; and press the sapphire crystal at 6:00 in order to reset the chronograph hands to zero. To achieve this spectacular feat, Jaeger-LeCoultre has developed a case composed of several perfectly integrated parts honed to within micron-level tolerances. And within this overall sculptural aesthetic appearance, the engineers have succeeded in recessing correctors into the material itself by means of intermediate wheels and connecting rods. To maximize the technical aspects of the moving parts distinguished by their powerful graphic design, the latter are partially visible through various dial apertures. All of which go to show that, for Jaeger-LeCoultre, practical purpose goes hand in hand with pleasing aesthetics.

*Jean-Christophe Babin, CEO, TAG Heuer.*

*FACING PAGE*
*Center • The Calibre 360 Concept Chronograph in its pink-gold version.*

*Botton • Toothed drive belts replace the traditional toothed wheels in the Monaco V4.*

## CEO of TAG Heuer

# Jean-Christophe Babin

**The philosophy driving the TAG Heuer research and development department is pretty straightforward. Watchmaking creativity depends on men and women's capacity to imagine shapes and volumes, but must also be governed by a certain number of constraints and specific features in order to give rise to a watch.**

## "Research and Development: Daring to take up a challenge with high added value."

At the end of the day, the finished product sums up a working charter including various intangible factors inherent to the TAG Heuer identity. "This is a task that involves certain compulsory parameters and that requires those conceiving it to show great open-mindedness and a critical view of the world," says TAG Heuer CEO Jean-Christophe Babin.

Almost 75% of the 25 people in this department work on products destined for series production. "Nonetheless," says Babin, "so as not to restrain their talents and limit their creativity, they also spend time in the unit dealing with the development of new concepts. The innate richness of the products stems from such multiple encounters and exchanges of ideas." According to Babin, this also "helps to ensure that each TAG Heuer watch combines the values of prestige and performance, sport and glamour that constitute the very bedrock of the brand and its unmistakable DNA."

### A distinctive financial set-up

"At TAG Heuer, the money allocated to the R&D department is thoroughly spent," admits Babin. To make sure there is no possible ambiguity in the perception of his words, he adds, "We consider the sum annually invested in this side of the company as lost. That eliminates any notions of profitability which may in time prove prejudi-

TAG HEUER
VANQUISH
CALIBRE 360
CHRONOGRAPH
1/100th
1/100
SWISS MADE

*FACING PAGE*
*The Vanquish version of Calibre 360 incorporates several Formula 1 type materials such as grade-5 titanium.*

*THIS PAGE*
*The Calibre S by TAG Heuer offers a new take on the quartz analog chronograph.*

cial to the life of a product." Acting in this way alleviates the temptation to launch a watch prematurely in order to recoup the costs involved in its construction. Adopting such an approach calls for a certain degree of boldness, but as the TAG Heuer CEO points out, "when the ever fascinating creative exercise is conducted independently of any call for profitability, the result is far more uninhibited and often leads to excellent results. Finally, believing in R&D is about daring to take up a challenge with high added value." Thanks to this distinctive approach, the brand has come up with such resolutely innovative products as the Monaco Sixty Nine, Monaco V4, Calibre 360 and Calibre S.

### Concepts currently in the pipeline

The public tends to believe that "concept watches" are placed for sale in the same year they are presented. However, Babin reminds everyone "that it may take years before TAG Heuer succeeds in producing innovative and reliable timepieces in quantities liable to meet market demand for the brand." Once a watch is presented, considerable thought must be given to the sturdiness of the mechanism and to the effectiveness of functions such as the chronograph reset mechanism. In the case of the Calibre S powering the Monaco V4, or the Chronograph Calibre 360, certain unexpected strains to which the movement is subjected were discovered during the pre-industrialization phases. Rapid rotations (Calibre 360), the transmission of forces via a driving belt (like in the V4) or the bidirectional nature of the readoff mode (Calibre S) have forced TAG Heuer to rethink certain constructions, but especially to redesign the profiles of all teeth in the gear trains of the models being considered in order to comply with NIHS norms. Reaching the point of preparation for large series production calls for considerable investments, but the final returns on such a policy are significant. Just as the innovations applied to Formula One motor racing often find an application in series-made cars, the discoveries made for watches should in the near future significantly enhance the efficiency of series-made watch instruments such as the Sixty Nine and the Calibre 360 chronograph currently in production; the Calibre S models currently in the industrialization phase; and the Monaco V4 (on the verge of completion).

# CEO Hublot

# Jean-Claude Biver

## Bigger Bang tourbillon column-wheel chronograph

"The Bigger Bang has two rare, distinctive features, one of them totally unprecedented. Firstly, it is a chronograph equipped with a flying tourbillon; secondly, the movement's column wheel has been turned to make it visible, representing a technical tour de force," explains Hublot CEO Jean-Claude Biver.

"Hublot lives on just one thing: Innovation."

"We set ourselves the task of adding a complication to a chronograph function. The tourbillon was chosen from among several other options, such as the perpetual calendar or the minute repeater." Complex to assemble, technically sophisticated, and an eloquent metaphor of perpetual motion, the tourbillon had every reason to be selected in that it vividly expresses the futuristic and sporty spirit of this watch.

Hublot embodies the crossroads between two worlds, a window in space and time located precisely on the frontier between past and future, tradition and innovation. In 1980, Carlo Crocco founded the brand and thereby established its DNA that may be summed up in just one word: Fusion. This creative mind was the first to fit rubber straps on gold cases. Within the context of that period, this move was viewed as an unacceptable revolution and its author was duly dubbed a "heretic." But the idea caught people's attention and an audience reaching to the head of the Spanish monarchy. HM King Juan Carlos of Spain has indeed recently offered Hublot models to his entire entourage. These days, it is rare in the sports watch category for any models to be offered without the option of a rubber strap. Since his appointment as CEO at the end of 2004, Biver has focused on instilling a new taste for success into the formerly somewhat dormant brand.

*FACING PAGE*
*Jean-Claude Biver, CEO of Hublot: "Hublot draws its strength from its capacity for innovation."*

*THIS PAGE*
*A detail of the Bigger Bang's movement.*

"Hublot's advantage lies in its extremely clear-cut concept. Fusion, meaning the passion for unusual unions are like a highway; once you're on it, you can't get lost," comments the dynamic Biver. "My role has been to rediscover and revitalize the DNA of Hublot, and to turn it into a religion with its own precepts. Once this is done, the rest follows quite naturally."

The conquest of atypical associations keeps Biver's team perpetually on the alert, tirelessly scouting out new and unexpected developments. Keeping this unique kind of innovation watch has certainly already borne fruit. Among timepiece connoisseurs, who could have missed the All Black icon model that won the 2005 Geneva Watchmaking Grand Prix in the design category? As its name implies, the All Black is entirely clad in black, while its ceramic dial confirms the brand's self-granted freedom to incorporate and combine unexpected materials. Its sharp lines and matte look literally set the market on fire and the models were snapped up within no time.

*Top • The Bigger Bang with platinum case and black ceramic bezel.*

*Top right • The Bigger Bang's movement has a visible column wheel.*

*Facing page • A red-gold version of the Bigger Bang.*

With the Bigger Bang, Hublot is innovating in spectacular fashion. First of all, through the aesthetics of the watch, enabling one to admire the entire system through a transparent sapphire crystal dial. Meanwhile, the indispensable indications

are provided by several finely numbered and marked circles. But above all, the Bigger Bang reveals a key component in the chronograph function, the column wheel. Normally placed on the back of the watch, turning it over called for considerable efforts by Hublot's Research and Development team. Located at 12:00, it further emphasizes the powerful mechanical appearance of the watch. The system is triggered by a single pushpiece that sets the column-wheel mechanism into motion, and reset by pressing the same pushpiece that stops the mechanism by the rotating motion of the coupling wheel engaging with the tourbillon system.

In addition to its visible column wheel, the Bigger Bang confirms its stature as an exceptional model by a 13-ligne tourbillon—measuring 30mm in diameter—with a 13mm-diameter carriage. The impressively sized flying tourbillon is fitted without a ball-bearing mechanism and spins 2.8mm above the under side of the case. This height creates the impression that the tourbillon is literally flying inside its carriage without any visible point of attachment.

Each month, the Hublot workshops produce around ten of these watches. But Biver aims to raise the bar and expects to be able to produce up to 15 a month as soon as possible, starting by meeting the dozens of orders already placed. With the Bigger Bang, the Nyon-based firm reveals the heart of a unique chronograph in which the contemporary watchmaking art has perfected a talented new interpretation of a well-established traditional complication.

CEO of Zenith

# Thierry Nataf

## El Primero chronograph movements

Just as spectacular as a magician pulling a rabbit out of a hat—and sometimes more so—Thierry Nataf, the charismatic and glamorous CEO of Zenith, excels in the art of surprising his audience by the sheer audacity of his creations. Since Nataf took the helm of this flagship brand in the small fleet of true chronograph manufacturers, in 2001, Zenith has undergone a quantum leap.

"I like short times that last."

Right from the start, Nataf sensed Zenith's phenomenal potential waiting only to be revealed. This treasure was a history driven right from the outset by the pursuit of ever more accurate time measurement. When Georges Favre-Jacot founded Zenith in 1865, he could scarcely have imagined the exceptional destiny in store for his company. In the 20th century alone, the manufacture would create over 50 legendary calibers and win 1,565 awards. But its crowning achievement came in 1969 with the presentation of the El Primero movement, the first ever to combine the chronograph complication with a self-winding mechanism. Ever since, the manufacture has constantly fine-tuned its horological gem, gradually forging the definitive symbol of the brand. Several decades later, in 1994, the ultra-thin Elite caliber was added to Zenith's exclusive range of exceptional proprietary movements.

The introduction of the El Primero—its name comes from Esperanto, an origin that expresses the universal nature of the humanist tradition—heralded a new era in the ultra-precise measurement of time cherished by the brand. It was the first movement to ensure measurement to the nearest 10th of a second. El Primero also featured a higher number of vibrations per hour (36,000) than any of its competitors, and therefore the highest degree of

*FACING PAGE*
*Thierry Nataf, CEO, Zenith.*

*THIS PAGE*
*Defy Xtreme watches ensure total water -resistance to a depth of 1,000 meters.*

accuracy. Upon his arrival, Nataf immediately grasped the benefits that could be derived from such a splendid invention and decided to "open up the secret," as he put it, in order to unveil this superb mechanism. This resulted in the immediate launch of the Open collection, distinguished by a special cut-out zone on the dial revealing the beating heart of the watch. Among all 18 variations of the collection, the Grande ChronoMaster Open El Primero is undoubtedly the most accomplished. Nataf then addressed a feminine audience, creating in its honor the Star collection, in which diamonds and precious metals vie for the spotlight. Here, too, the dial reveals the insides of these glamorous and sensual watches.

After Nataf had explored tradition and completely revamped the classic models, all that was left to achieve was a resolutely contemporary interpretation of the El Primero movement. In 2006, the brand created a sensation by presenting two new sports lines: the Defy Xtreme and the Defy Classic. Representing genuine technical fortresses, the Defy watches confirmed the era of the absolute watch—from its sheer size to its movement and materials. Specially for these models, the El Primero calibers have been transformed into SC and SX versions in which the bridges supporting the mechanism are in Zenithium, an alloy developed by the manufacture that combines titanium,

*THIS PAGE*
*Top right • A Zenith pocket -watch.*

*Above • The models in the Open range reveal the El Primero movement beating at 36,000 vibrations per hour, as on in the ChronoMaster Open Neo-Vintage model pictured here.*

*FACING PAGE*
*Left • Zenith pays tribute to women with its Star collection.*

*Right • The brand cultivates an unconventional approach to design and enjoys exploring unexpected worlds such as glam-rock.*

aluminum and nobium. The dials feature the use of aluminum honeycomb, carbon fiber and hesalite materials.

Virile, radical, superlative—and fashionable—the Defy Xtreme watches are water resistant to 1,000 meters. Crafted in ultra-trendy blackened steel on a steel bracelet, these models, measuring 43mm for the Elite series and 46.5mm for the chronographs or the tourbillon, are emblems of a certain outsized extravagance. The Defy Classic range, a touch more restrained in its presentation and a tad more reasonable in its design, meets the expectations of a public looking for luxury timepieces, yet prepared for adventure on a daily basis. While its general case dimensions are identical to the Xtreme versions, the Classic line is made not from titanium but from high-grade steel. Its water resistance is guaranteed to a depth of 300 meters.

A proven master in the art of springing surprises, Nataf has announced the 2007 introduction of a high-speed chronograph. "We are working on a high-speed caliber with extreme performances topping 36,000 vibrations per hour. Our goal is to reach the domain of the ultra-precise. This will involve parameters such as a revolution in the regulating organ," explains Nataf. And for the future, he issues advance warning: We can expect a full-scale horological cataclysm for 2010-2011. The epicenter of this earth-shaking event is already known, since he is, of course, referring to Zenith.

De Bethune

# David Zanetta and Denis Flageollet

## The Maxichrono

"I'm all in favor of perpetual evolution. I prefer that rather than maintaining time-honored traditions that are no longer relevant," says Technical Director Denis Flageollet, who makes no secret of his fascination for innovation. Within five years, he has actually achieved what many other brands could only dream of, namely creating four futuristic and innovative calibers from scratch.

"De Bethune is moving steadily towards the watch object by achieving perfect integration of the case and the movement."

The Maxichrono, based on a concept presented in 2006 and that will be shown in prototype form at the spring watch fairs in 2007, is the latest in this series. Within a short space of time, De Bethune has demonstrated a powerful capacity for innovation. Witness its all-silicon thermo-compensated sprung balance. More reliable and more accurate than conventional balances, it paves the way for "intelligent components," or "smart parts." Faced with an increase in temperature, the silicon balance-spring reacts by slowing down, which is why in the same circumstances the De Bethune changes shape and the platinum inertia-blocks move towards the center, making the balance pick up speed and thereby compensating for the tardiness of the balance-spring. Moreover, this innovation incorporates all of the characteristic properties of silicon. Representing a major new invention by the brand, it is compatible with all models and the Research and Development department is constantly working to improve the mechanism.

With its single pusher operating five co-axial hands, the Maxichrono conceals several exceptional technical innovations beneath its understated exterior. The first is the triple shock-absorbing system created by the brand itself and destined to protect the original sprung

*FACING PAGE*
*Top left*
*David Zanetta,*
*CEO of De Bethune.*

*Right*
*Denis Flageollet, Technical Director for De Bethune.*

*THIS PAGE*
*Top*
*The Maxichrono seen from the back.*

*Above*
*The Maxichrono mono-pusher watch developed by the house of De Bethune.*

balance system. It exists in a titanium balance-bridge to which two jewel endstones are held by a double spring. The five hands of the Maxichrono are superimposed in the dial center, based on a design specially created for this model. This technical feat relates to a desire to improve the readability of the chronograph at all times, independently of the position of the hour and minute hands. The latter, when appearing on the traditional tiny counters, can be extremely hard to read. On this model, each has its own reading scale virtually equivalent to the size of the dial itself. The Maxichrono is also distinguished from virtually all its com-

*Top left • DBS, the first watch from De Bethune, was already a gem of technical innovation.*

*Top right • The Beat sports watch from De Bethune has central power reserve.*

*Above • The feminine counterpart of the DBL has a patented moonphase system.*

*Facing page • A wintertime setting for the De Bethune workshops in La Chaux L'Auberson.*

petitors by its orange chronograph seconds hands that makes a complete rotation in 30 seconds instead of the customary 60, rendering read-off both faster and easier. Meanwhile, the scale of the minute counter indicates both the minutes and half-minutes elapsed since the start of the timing operation. In order to reach the goal that was established when this caliber was first conceived, the very operation of the chronograph itself had to be redesigned. De Bethune's Research and Development team decided to separate the functions of the three counters—seconds, minutes and hours—thereby leading to the use of three column wheels instead of the customary one.

This caliber has been entirely developed and crafted by De Bethune. Equipped with a self-regulating twin barrel, it accumulates more than 120 hours of power reserve. The off-centered coupling-clutch system is totally innovative and enables instant and ultra-precise operation of the seconds hand. The hour and minute hands move forward gradually thanks to a continuous movement.

In terms of aesthetics, De Bethune watches aim to achieve total harmony. As brand CEO David Zanetta explained when the company was founded in 2002, "Watches are probably the objects that have been most closely studied and designed for over three centuries; they have been improved millions of times to the point of reaching their quintessence in our day—especially since its basic round shape constitutes the only basis for progress."

In the case of the Maxichrono, the design of the watch stems from its technical characteristics. The multi-colored hour and minutes hands, which require meticulous care to produce, are made in-house. The generously sized dial comprises three levels, each corresponding to a different function. The colors alternate between a matte silvered shade and the gray of the titanium circles eliminates any unpleasant glinting effects. An antireflective sapphire crystal with special double treatment protects the dial and enables optimal readability of all functions, whatever the lighting conditions. The ergonomic case of the Maxichrono is fitted with pivoting shafts that adjust to the wearer's wrist. It thereby underscores the lines of its caseband and facilitates the use of the mono-pusher at 6:00. As with the hands, the 27 parts composing the case are made exclusively within the manufacture. The last fine-tuning was performed in 2006 and the Maxichrono debuted in its definitive version at BaselWorld 2007.

# CEO of Richard Mille

# Richard Mille

## RM - 011

When launching his brand, Richard Mille would have had a tough time convincing informed connoisseurs to buy his models if they had not been eloquent in themselves. At a time when haute horology was barely rising from the ashes after the traumatic bombshell arrival of quartz technology, the industry was more concerned with reviving a prestigious past and glorifying tradition than with implementing aerospace technology.

"The important thing is to explain the tremendous difficulties involved in making these models, as well as the horological philosophy and the beauty they embody."

—Richard Mille, explaining how he helps people grasp his approach



Today more than ever, Mille loves futuristic materials, Formula 1 motor racing, and crazy challenges. And collectors are coming along for the ride!

From his magnificent retreat in Brittany, Mille handles his brand as one would drive a mean machine on a racing track: very fast. Those words are indeed ones he lives by. However, the comparison extends to his accurate and efficient driving, with no fuss or frills. Mille began by working for others: the Matra watchmaking division, the Compagnie Générale Horlogère (Seiko Group), Mauboussin, Baccarat, Repossi, as well as cooperating with Audemars Piguet. These experiences sharpened his knowledge of the watchmaking world and enabled him to fine-tune his own marketing vision.

His determination to reach the peak of watchmaking quality has led him to work with the finest movement specialists right from the start. From Audemars Piguet Renaud Papi (APRP) to the Vaucher Manufacture Fleurier, Mille knows how to enlist expert assistance in creating revolutionary movements, which double as perfect metaphors for car engines—all from scratch. At the turn of the century, the watchmaker put a stop to nasty rumors regarding his "dangerous" ideas (which some thought would doom Mille to failure) by presenting the RM 001. This entirely new breed of tourbillon was a sensational surprise and won the

*FACING PAGE*
*Richard Mille, CEO, Richard Mille.*

*THIS PAGE*
*Back of the ultra-light RM 009.*

THIS PAGE
The RM 011 is fitted with an adjustable oscillating weight.

FACING PAGE
Left • Back of the RM 011 chronograph.

Right • Richard Mille has used alusic, a material from the aerospace industry, to make the caseband of the RM 009, a watch weighing just 30 grams.

steadfast devotion of many collectors. Since then we have witnessed a succession of new models presented at a rate of one, two or even three per year. Eager for novelty but dedicated to perfection, Richard Mille constantly improves upon existing models. In the artist's own words, "it's a constantly evolving and ongoing process." One can thus find RM 001 watches as well as RM 001 V2 models, which represent the second version of the timepiece. Indeed, this particular model has already undergone 16 different evolutions. Much like a rugby team, Richard Mille moves ahead in a close-knit formation, setting new records along the way, including that of the world's lightest watch (weighing a mere 30 grams without the wristband), named RM 009FM, in tribute to Brazilian race-car driver Felipe Massa, a fervent devotee of the brand.

### The RM 011 chronograph

The RM 011 will be unveiled to the public for the first time this year, filling a lacuna in the brand range. None of the current models bridge the gap between the simplest models and the incredibly complicated ones—a role that will be played by the RM 011, which personifies the Richard Mille spirit with its transparent sapphire crystal dial, revealing the whole movement. Only the white numerals, designed by the watchmaker himself and representing one of the brand's many signature details, appear above the gear wheels. The large date window positioned below 12:00 highlights the day numeral with a striped red band. The large tachometer scale indexes mark out the rim of the dial. Both the pushers and the crown have been the object of particular attention, adding an elegant touch to the tonneau-shaped case, which is framed by 12 star-shaped screws that were developed by Mille himself. The crown evokes the wheel of a Formula 1 race-car wheel with its hollowed surface reinforced by concentric stripes. To complete the effect, it is clad in a soft black rubber strip that enhances the grip. Meanwhile, the edges of the lengthwise-striped pushers are engraved in red with "start/stop" and "reset/flyback" indications. As a whole, the RM 011 radiates a sense of raw power that confirms the parallels with high-tech vehicles roaring around the racing circuits. The perfect finishing and the meticulous concern for detail transform this timepiece into an object worthy of lengthy contemplation. The intricate maze of gear wheels is itself enough to nurture such futuristic meditation. Mille intends to produce several hundred RM 011 watches per year. But will that be enough to meet the exponential demand? While the watchmaker definitely loves speed, he also believes in instilling patience. "When I explain to customers that watches are not industrial products and that deadlines sometimes simply cannot be met, they understand the situation perfectly and are prepared to wait," says Mille.

CEO
of Maurice
Lacroix

# Philip C. Merk

## Maurice Lacroix ML 106 caliber Chronograph

Maurice Lacroix had received a boost in popularity thanks to its first ambassador, Roger Federer. On the advice of his managers, however, the tennis champion bought back the contract tying him to the manufacture. In a surprising paradox, the end of this cooperation proved a great thing for the brand, based in the Franches-Montagnes region.

"The chronograph is still the most subtle complication of all."

"Roger Federer was a perfect ambassador," says Maurice Lacroix CEO Philippe C. Merk. "We have earned a great deal of brand awareness thanks to him. The end of this partnership enables us to build a new platform for the brand, in which the notion of a manufacture takes first place. We want to build a new brand image, and tennis would probably not have been the best vehicle for that."

The ML 106 caliber is the first manufacture-made movement from Maurice Lacroix. It is a traditional chronograph movement with a column-wheel and swan-neck precision adjustment. For this chronograph, Maurice Lacroix has reinterpreted the chronograph's cinematic chain. The main innovation is the use of a release lever. This is positioned under the hammer and comes into contact with the blocker during the zeroing process. On traditional chronographs, the chronograph wheel is free for fractions of a second during the "start" and "reset" sequences. If an impact occurs, this may result in incorrect displays. This new special mechanism minimizes such a risk; it guarantees the correct positioning of the individual chronograph parts and hence the precision of the displayed information. A patent is pending for this new interpretation. The chronograph parts are polished and adjusted by hand in the Maurice Lacroix workshops.

*FACING PAGE*
*Philippe C. Merk, CEO of Maurice Lacroix says, "We are progressively nearing our objective: that of becoming a full-fledged manufacture."*

*THIS PAGE*
*Top*
*Maurice Lacroix intends to distinguish itself with large models.*

*Above*
*The Maurice Lacroix buildings in the Jura region of Saignelégier.*

*Top left • ML3. Details of the ML 106 chronograph movement.*

*Top right • The first in-house Maurice Lacroix chronograph movement is the ML 106 caliber.*

*Facing page • Within the field of mechanical watches, Maurice Lacroix has also distinguished itself with retrograde displays.*

Another function of the release lever is to lock the zeroing function when the chronograph is engaged, rendering procedural errors impossible since the chronograph must be stopped before it can be reset to zero. A special feature of this chronograph is also the 60-minute counter.

The imposing 45mm ML 106 chronograph pays tribute to the period when the first chronographs appeared on the scene. "The chronographs on the market today are often too small and the details are becoming illegible. We have thus decided to think big in order make our models more readable," explains Merk. "While it is true that larger models are enjoying considerable success at the moment, we are not following fashion but have made this choice for aesthetic reasons. In fact, most of our models are fairly large," he adds. By remaining loyal to and reproducing early models, Maurice Lacroix proudly proclaims its passion for these early versions. The ML 106 chronograph thus beats at 18,000 vibrations per hour—a fairly slow cadence, corresponding to that of the first chronographs.

Maurice Lacroix has made the momentous decision to produce movements in-house. "It is a major step to move from producing modules, as we did before, to making complete mechanisms," comments Merk. The firm is thus pursuing a strategic option devised several years ago. Whereas it was offering mainly quartz-driven products at the time, the company fundamentally steered its strategy toward the direction of mechanical horology in the late 1990s. As a preliminary step, Maurice Lacroix had integrated part of its production in 1989 by buying up the Queloz case manufacturer, also based in Saignelégier. Including its four subsidiaries (in Switzerland, Germany, the United States and the United Kingdom), the company now employs 220 people and achieves a 100 million Swiss franc turnover, of which 60% stems from mechanical models driven by the Masterpiece collection. In addition to the complication modules developed in-house, and while maintaining its chronograph movement, the company is making the most of the variety currently available, as it sources its movements both from ETA and from newcomers to the mechanical watch market such as Sellita and Technotime. The move upscale has obviously been to the detriment of production volume (currently standing at 150,000 units per year) and the repositioning will naturally entail some difficult years. Germany is still the leading market of the brand, which achieves 60% of its turnover in Europe, followed by Asia with 23% and the United States with 18%.

# Chronographs, Chronograph Rattrapantes and Flyback

## Close-up on chronographs

Among the major trends that have characterized the most recent watchmaking vintages, one cannot ignore the dramatic rise in the number and variety of chronographs. Whether in the shape of "simple" mechanisms or split-second chronographs, this complication is enjoying unprecedented success, with mechanical chronographs taking a premier position.

**There are several reasons behind this phenomenon.**

First, the chronograph is the perfect emblem of modern times. Born in the 19th century with the industrial boom and the surge of competitive sports, does it not eloquently embody our passion for speed, precision and performance, as well as our taste for mastery and measurable accomplishments?

Second, its many practical applications make it one of the most useful complications in everyday life—although many buyers of this type of watch choose it more for the aesthetic appearance of the dial than for actually timing records!

And third, of course, is that very aesthetic appeal with a technical touch so closely attuned to the growing wave of sporty chic timepieces, making these both elegant and casual watches that are equally at home in the office, on the tennis court or at an evening event. In short, they are versatile accessories in tune with a new lifestyle.

Chronographs definitely have the wind in their sails at the moment. Fostered by the advancement of computerized design and production technologies and boosted by public interest, watchmakers are innovating on all fronts, meaning in terms of mechanisms, materials and designs——and often all three at once. Driven by consumers seeking exclusivity, several brands have recently launched their own proprietary chronograph movement, generally enhanced by innovative technical features. As for the split-second chronograph, which is making noteworthy progress, it has retained and indeed reinforced its status as one of the most sophisticated and prestigious complications in the art of horology.

Another sign of the times is the fact that chronographs, long considered an exclusively masculine preserve, are increasingly appealing to women. Rather than merely having to borrow their companion's timepiece, they are being gratified with an increasing number of models destined specifically for their feminine wrists. And if, as the popular French song goes, "woman is the future of man," she is also undoubtedly that of watchmaking——and manufacturers have been quick to grasp the importance of this evolution.

In the 1970s, the quartz tidal wave dealt a near-death blow to mechanical watchmaking, and particularly to the chronograph, which could not hope to vie with quartz in terms of precision. However, the flurry of new models appearing at the dawn of this third millennium and the promising dynamism shown by watchmakers in this field prove that mechanical chronographs have lost none of their aura nor of their power of seduction among connoisseurs, and are indeed gearing up to enjoy a new heyday.

# Chronographs

The term "chronograph" refers to a watch endowed with an additional mechanism serving to measure and display elapsed times. A faithful companion of sports enthusiasts, scientists and all those fascinated by the mastery of time, this extremely useful complication is currently surfing an unprecedented wave of popularity. It has found its way into dressy evening events, shows up in both town and country, and is setting off to light up feminine wrists. However, the rapidly expanding number of chronographs should not obscure the fact that this is a genuinely sophisticated mechanism, especially in its noblest form: the column-wheel chronograph. Indeed, the many technical and aesthetic innovations recently offered by watchmakers confirm that this field still has plenty of fine surprises in store for devotees of mechanical watches.

### The embodiment of modern times

Of all the complications in the classic repertoire„" the chronograph is one of the most recent. Before being able to measure short times, watches first had to reach a sufficient level of precision. Yet the minute hand did not appear until after 1675 (when the balance-spring was invented by Huygens) and it was only in the latter half of the 18th century that dials began to feature a seconds hand. Watchmakers also had to devise a system capable of immobilizing the seconds hand in order to perform measurement. The initial solution consisted of stopping the movement, as with the "deadbeat seconds" watch by Romilly (1754), described in the Encyclopedia of Diderot and d'Alembert, a timepiece fitted with a central hand performing one jump per second. In 1776, Geneva watchmaker Jean-Moïse Pouazit invented the "independent deadbeat seconds" watch. Equipped with a second gear-train, the direct-drive seconds hand could be started and stopped at will without disturbing the running of the movement. Nonetheless, this type of watch still required the user to write down the start of the measurement and to perform a calculation, since the seconds hand could not be brought back to zero.

1

2

▸▸ 1 Breitling, Navitimer
*With its circular slide-rule, the Navitimer by Breitling immediately asserted itself as a cult watch for pilots and aviation enthusiasts.*

▸▸ 2 Omega, Speedmaster
*The Omega Speedmaster had the privilege of accompanying the crew of the Apollo 11 space shuttle in July 1969, earning it the "Moonwatch" nickname.*

The first chronograph truly worthy of the name dates back to 1822 and bears the signature of Nicolas Mathieu Rieussec——unless it was in fact an invention by the famous Abraham-Louis Breguet (an issue that is the object of a fierce dispute). It is distinguished by an original display mode: to indicate the time to the nearest 1/5th of a second, the hand deposits a drop of ink on the dial—hence its name drawn from the Greek words chronos, meaning time, and grapho, to write. Towards the mid 19th century, a watchmaker from the Vallée de Joux, Adolphe Nicole, registered a patent for the first ever reset or return-to-zero system. Thus equipped with its three essential functions (start, stop and reset) the chronograph quickly proved its worth in countless areas (science, technologies, transport, the military, etc.). The function became the perfect emblem of modern times and it is no coincidence that its expansion coincided with industrialization and one of its spin-offs: competitive sports. The chronograph was also progressively enriched with various types of graduated scales serving to perform various measurements: the tachometer (speed), the telemeter (distance), the pulsometer (pulse rate), etc.

## From the pocket to the wrist

The first wrist-worn chronographs appeared in 1915 and they rapidly earned considerable success with the public and among specialized users. Like pocket chronographs, they were equipped with a mono-pusher for all three functions, generally housed in the crown. In 1934, it was Breitling who gave the chronograph its current configuration by creating the second reset pusher. In 1937, the Dubois Dépraz movement manufacturer presented a movement in which the chronograph was controlled by a cam system that was less expensive and less tricky to make than the traditional column-wheel mechanism. This system was gradually adopted by most brands. The chronograph pursued its glorious career with the birth of several great classics. 1952 saw the launch of the **Breitling Navitimer**, with its inimitable circular

▸▸ 3 *Breitling, Chrono-Matic (réédition de 2004)*
*Breitling's Chrono-Matic, the first self-winding chronograph watch (1969), is shown here in its 2004 re-edition. Note the left-hand crown.*

▸▸ 4 *Rolex, Cosmograph Daytona*
*The updated version of the legendary Cosmograph Daytona houses a proprietary Rolex chronograph movement.*

slide-rule serving to perform all calculations relating to airborne navigation——a cult object that has been in uninterrupted production for more than 50 years. Moving into an even higher (strato)sphere, the Speedmaster by Omega accompanied American astronauts on their first lunar mission, earning it the famous "Moonwatch" nickname.

## The birth of the self-winding chronograph

1969 actually a second small step for man—but a giant step for watchmaking: The creation of the first, or rather the two first, self-winding chronograph movements. While there is still heated debate as to which was "announced" and which was "presented" first, both movements are clearly based on a somewhat different construction.

The one developed by the Breitling/Büren/Heuer-Leonidas trio is a modular-built cam movement with off-centered microrotor (oscillating weight). It is the forerunner of most self-winding chronographs used today. For Breitling, this Caliber 11 would lead to the development of an entire range of chronographs named Chrono-Matic, recognizable by their left-hand crown. In 2004, the brand paid tribute to the original Chrono-Matic by offering not a mere re-edition, but rather a reinterpretation of this model in a decidedly contemporary spirit.

As for the caliber presented by Zenith in 1969, the famous El Primero is an integrated movement with column-wheel control and a central rotor. It was also the first mechanical chronograph movement capable of measuring times to within 1/10$^{th}$ of a second. In addition to Zenith models, the El Primero movement was to equip certain watches by some of the greatest names in the watch business. For instance, El Primero could be found at the heart of the legendary Cosmograph Daytona before Rolex developed its own chronograph movement at the turn of the

1

21st century. Zenith, currently in the midst of a spectacular revival, has recently restored its finest historical treasure to its rightful position by creating models such as the Grande ChronoMaster Open El Primero featuring a dial opening enabling one to admire the beating heart of the watch.

It is worth noting that there are very few brands who actually make their own movements and that most of the mechanical chronographs offered under various labels house an ETA Valjoux 7750 movement, remodeled to varying degrees and sold at widely differing prices.

1 *Zenith, Grande ChronoMaster Open El Primero*
*The Zenith Grande ChronoMaster Open El Primero's dial aperture reveals its famous movement beating at 36,000 vibrations per hour.*

### The quest for exclusivity

Today, things are beginning to change. Buyers are increasingly informed and skilled connoisseurs are prepared to pay the price

▸▸ 2 *Patek Philippe, Annual Calendar Chronograph Réf. 5960P*
*Patek Philippe's Annual Calendar Chronograph is distinguished by its vertical coupling-clutch and its large minute/hour counter.*

▸▸ 3 *Chopard, L.U.C Chrono One Concept Watch*
*Celebrating the 10th anniversary of Chopard Manufacture's production of the L.U.C calibers, the brand presented its first chronograph movement in a limited edition (2006).*

for a chronograph from a major brand provided it is driven by an exclusive (or at least somewhat original) movement. Brands have acknowledged this fact and the industry is seeing a spate of proprietary chronograph movements featuring innovative solutions. Such is the case for Patek Philippe. Whereas its famous chronographs had hitherto been equipped with movements based on a blank produced exclusively for Patek Philippe by Nouvelle Lémania, in 2006 the Geneva-based firm created a sensation by presenting its Annual Calendar Chronograph Ref. 5960P equipped with the first self-winding chronograph movement entirely conceived and developed in its own workshops. In keeping with its reputation, the manufacture has introduced several interesting innovations and distinctive features. Thanks to a disk clutch system operating virtually without friction, and thus involving minimum wear, the large direct-drive chronograph seconds hand may also be used as a continuously running seconds display. In terms of design, the Annual Calendar Chronograph Ref. 5960P is equipped with a patented "annual calendar" (see Perpetual Calendars chapter) and is distinguished by a large single counter combining both hour and minute totalizers.

In September 2006, Chopard celebrated the 10th anniversary of its L.U.C Manufacture by launching its first ever chronograph movement, L.U.C Caliber 10CF. Three patents are pending for this integrated self-winding column-wheel movement with vertical coupling-clutch, flyback function and 60-hour power reserve. The pusher at 4:00 serves to reset the small seconds, representing an extremely useful function for ensuring ultra-accurate time setting. Chronometer-certified by the COSC, this movement was unveiled as the driving force within the L.U.C Chrono One Concept Watch," which was issued in a limited series of 100.

2

3

1

▸▸ 1 *Maurice Lacroix, Masterpiece Le Chronographe*
*Maurice Lacroix confirmed its transition to upscale, high-end mechanical mastery with the Masterpiece Le Chronographe powered by Maurice Lacroix's first in-house movement.*

▸▸ 2 *F.P. Journe, Octa Chronograph*
*The Octa Chronograph from F.P. Journe is the first self-winding chronograph to maintain its precision for five days without being worn.*

▸▸ 3 *Jaeger-LeCoultre, Master Compressor Extreme World Chronograph*
*The Jaeger-LeCoultre Master Compressor Extreme World Chronograph features a case equipped with an unprecedented shock-absorbing system.*

▸▸ 4 *Jaeger-LeCoultre, AMVOX2 Chronograph Concept*
*A revolutionary idea from Jaeger-LeCoultre: Simply pressing the AMVOX 2 Chronograph's crystal Concept activates its controls.*

Maurice Lacroix has confirmed its upscale move towards mechanical high-end models by presenting its first proprietary movement housed within the Masterpiece Le Chronographe. This hand-wound column-wheel chronograph movement features an innovative lever system for the start and reset functions. Meanwhile, F.P. Journe has enriched its **Octa** collection with the first self-winding chronograph able to maintain its precision for five full days without being worn and graced with a regulator-type dial characteristic of the Geneva-based brand.

## A new gesture

In 2005, Jaeger-LeCoultre presented its first self-winding chronograph in a case equipped with an original shock-absorbing system: the Master Compressor Extreme World Chronograph. However, its most spectacular innovation in this field was introduced in 2006. Within the AMVOX2 Chronograph Concept, inspired by its partnership with Aston Martin, the manufacturer in Le Sentier has quite simply reinvented the manner of activating the chronograph controls: traditional pushers are replaced by a system based on a case middle that pivots slightly along the 9:00—3:00axis. To activate the chronograph, all the user need do is gently press the upper part of the sapphire crystal, and the same goes for stopping it, whereas pressing the lower part will reset the function. So as to avoid any inadvertent handling, a cursor or function selector placed along the case side at 9:00 :00locks the mechanism in three distinct positions. This revolutionary system, involving seamless interactivity between the watch exterior and movement, is housed within a titanium and steel case. The dial features counters reminiscent of the dashboard counters in the famous British Aston Martins.

## Ten times the accuracy

Since 1969, the precision record for mechanical chronographs was held by Zenith's famous El Primero, which is capable of

2

3

4

1

2

▸▸ 1 *TAG Heuer, New Aquaracer Calibre S*
*The New Aquaracer Calibre S by TAG Heuer displays the time, the timing, or the "regatta" countdown with three central hands and two patented bidirectional counters.*

▸▸ 2 *Glashütte Original, PanoRetroGraph*
*Glashütte Original's PanoRetroGraph is the first chronograph capable of counting backwards.*

▸▸ 3 *TAG Heuer, Carrera Calibre 360*
*TAG Heuer's Carrera Calibre 360 (in pink gold) is the first mechanical wristwatch with an accuracy to the nearest 1/100th of a second.*

measuring time to within 1/10th of a second. In 2005, TAG Heuer created a stir by presenting the Calibre 360 Concept Chronograph," the prototype of the first mechanical wrist chronograph accurate to within 1/100th of a second. This performance was achieved by combining two independent movements, the second of which (and the one driving the chronograph) beats at a frequency of 360,000 vibrations per hour. This avant-garde movement was offered in 2006 in a pink-gold, limited edition of the **Carrera**—a legendary chronograph born in 1964. TAG Heuer, which is obviously at peak form, has also been busy innovating in terms of display and presents the **New Aquaracer Calibre S.** This original watch is distinguished by its instant and universal read-off system—(which the brand boldly touts as the "easiest to read in the world")—based on three central hands and two counters with patented bidirectional hands. The central hands are used to display the classic hours, minute and seconds; the hours, minutes and seconds of measured time; or the remaining minutes and seconds of a regatta countdown. These three functions are selected by simply pressing the crown; when the countdown is complete, the watch automatically reverts to chronograph mode. The bidirectional hand of the right-hand counter displays 1/10ths of a second in "chronograph" mode. In the normal "time" mode, the date is indicated by the bidirectional hands of the twin tens/units counters. While on the topic of the countdown functions, one should mention that the German firm Glashütte Original had already made a splash with its **PanoRetroGraph**, the first chronograph to count time "in both directions." This timepiece equipped with a flyback function enables one to launch a new timing operation by pressing the pusher at 4:00 without needing to stop and reset the hands.

### A new take on geometry

For many years, chronographs tended to sport a rather classic look with three small counters or subdials (hour and minute

3

1

2

totalizers, small seconds) arranged in a perfectly symmetrical manner. This traditional appearance is currently being challenged by many new models seeking to reinvent the reading of time. Eberhard & Co set the tone by launching the **Chrono4**, characterized by four counters (minutes, hours, 24 hours and small seconds) lined up either horizontally or vertically. In its **SplitRock** model, Pierre DeRoche focused on a simple, direct and analog read-off by introducing a chronograph with three concentric hands (hour, minute and seconds). Its exclusive Dubois Dépraz caliber enables one to read off the measured time at a glance, just as one reads the time every day on a watch.

3

Gérald Genta, a well-known specialist of retrograde displays (see Retrogrades Mechanisms and Jump Hour chapter) combines four indicators of this type within its **Arena Chrono Quattro Retro**: retrograde minutes on a segment between 10 and 2:00, 30-minute retrograde counter at 3:00, 12-hour retrograde counter at 9:00, and double-pointer date indication at 6:00—complemented by a jumping hour display at 12:00. The **Porsche Design Indicator by Eterna** offers an unusual read-off that is designed to be simpler and more efficient than the traditional subdials. The hours (from 1 to 9) and the minutes (in tens and units) of measured time are displayed in mechanical "numerical" mode by means of rotating disks appearing through a twin aperture at 3:00, while the seconds tick by in a more conventional manner with a central direct-drive seconds hand. This movement, comprising almost 800 parts and equipped with four barrels, is housed within a sturdy titanium case with a design strongly inspired by the Porsche Carrera GT. As for deLaCour brand, it has distinguished itself—albeit more in aesthetic terms than technical—by presenting the **BICHRONO** with a distinctive extra-wide case housing two self-winding chronograph movements.

4

5

▸▸ 1 Eberhard & Co. Chrono 4
Eberhard & Co. reinvents the face of the chronograph on its Chrono 4 with four linear counters.

▸▸ 2 Porsche Design Indicator by Eterna
Porsche Design Indicator by Eterna, with a double hour and minute aperture for timing operations.

▸▸ 3 deLaCour, Bichrono
The BICHRONO by deLaCour combines two self-winding movements within its elongated case.

▸▸ 4 Pierre DeRoche, SplitRock
Simplified read-off: the SplitRock chronograph by Pierre DeRoche features three concentrically arranged hour, minute and seconds hands.

▸▸ 5 Gérald Genta, Arena Chrono Quattro Retro
The Arena Chrono Quattro Retro model from Gérald Genta features no fewer than four retrograde displays.

60
45
15
30
ZENITH
El Primero
High Pressure Resistant Chronograph
1000 M
Condition Use

1

## New materials and new looks

The world of chronographs has also been transformed recently by the appearance of new materials that considerably increase their robustness, reduce their weight and improve their performances, as well as challenging existing aesthetic codes in order to appeal to a clientele eager for originality. Zenith set off a mini shockwave in 2006 by presenting its new Defy Xtreme collection, inspired by the world of mechanical sports, yachting and aviation, and designed "For Xtreme condition use." Research focused mainly on materials, including the development of an exclusive ultra-light and ultra-resistant alloy christened Zenithium and used on the mainplate and bridges. The black titanium case is water resistant to 1,000 meters. The multi-layered dial combines honeycomb aluminum, anti-shock hesalite and carbon fiber. The **Defy Xtreme Open** also reveals a new development of the El Primero movement, with a heart protected by blades reminiscent of a turbine or a jet engine. Endowed with ruggedly powerful high-tech charm, this design is exploring new territory within the field of sporty chic watches.

As the self-proclaimed advocate of "fusion," Hublot naturally combines materials in its new Big Bang models. The brand pulled off an effective publicity stunt in 2006 with the **Big Bang "All Black"**. Distinguished by a black ceramic case and bezel, black dial, numerals, hour-markers, hands and date display, complete with a black rubber strap, this model geared more towards originality than readability (Hublot poetically refers to the "invisible visibility of time") is also fitted with an exclusive movement. Meanwhile, Hublot's Mag Bang chronograph combines a magnesium case with a movement equipped with titanium mainplate, bridges and oscillating weight. The resulting "feather" weight is 22 grams for the movement and 72 grams for the finished watch. Magnesium, titanium, ceramics, synthetic fiber, texalium and rubber—talk about fusion!

▸▸ 1 *Zenith, Defy Xtreme Open*
*Zenith's Defy Xtreme Open: Powerful technology and design "for Xtreme condition use."*

▸▸ 2 *Hublot, Big Bang "All Black"*
*Big Bang "All Black" by Hublot, or the "invisible visibility of time."*

2

1

▸▸ 1 *Audemars Piguet, MC12 Tourbillon and Chronograph*
*Audemars Piguet has drawn inspiration from the automobile world for its MC12 Tourbillon and Chronograph with a double barrel ensuring a 10-day power reserve.*

▸▸ 2 *Zenith, Baby Doll Star Open Sea*
*Baby Doll Star Open Sea demonstrates Zenith's approach to ladies' chronographs.*

▸▸ 3 *Blancpain, Villeret Single Push-piece Chronographe*
*On the Villeret Single Push-button Chronograph by Blancpain, one pusher serves to operate the start, stop and reset functions.*

2

Among automobile-inspired models (the imagination of sports watch and chronograph design engineers is greatly stimulated by fine mechanisms on four wheels), Audemars Piguet paid tribute to Maserati and to one of its finest racing cars by creating the **MC12 Tourbillon Chronograph** endowed with a double barrel ensuring a 10-day power reserve. This original timepiece features an array of high-tech materials including a carbon-fiber mainplate and eloxed aluminum bridges as well as a bold technical look that includes an oval case, displays offset as if disrupted by a sudden burst of speed, and an openworked dial revealing the movement. The **MC12 Tourbillon Chronograph** was issued in a limited edition of 100 in platinum.

### Time for women

Another strong trend at the start of the 21st century was the rush of new chronograph models destined exclusively for women. Gone are the days when feminine sports watch lovers had no other choice than to wear a man's watch. Chronographs are now adopting soft colors, finely crafted dials and even diamonds, just like more inherently dressy watches. And more and more women are appreciating the sophistication of a finely crafted mechanical movement. Zenith had already set the tone with its Star El Primero models, including variations in dusty pink, candy blue and lime green, on lizard skin straps, with guilloché dials and diamond-set bezels. With the **Baby Doll Star Open Sea**, also

3

▸▸ 4 *Girard-Perregaux, Column-Wheel Chronograph Lady Team*
*The small column-wheel chronograph Lady Team Watch by Girard-Perregaux pays tribute to the feminine members of the BMW ORACLE crew in the America's Cup 2007.*

▸▸ 5 *Hublot, Big Bang Porto Cervo*
*The Big Bang Porto Cervo by Hublot embodies the "fusion" of red gold, white rubber and diamonds.*

equipped with the legendary El Primero movement, the Manufacturer in Le Locle offers a beautiful poetic digression on the marriage of sky and sea: mother-of-pearl dial with engraved star motif, dancing numerals and a pearl at 3:00; starfish-shaped opening revealing the heart of the movement; star-engraved bezel, stingray strap and several diamond-set versions.

4

Girard-Perregaux also offers a beautiful blend of technical and esthetic features within its self-winding ladies' Small Column-Wheel Chronograph. In 2006, the firm even developed an exclusive model dedicated to female members of the BMW ORACLE team it is supporting as Challenger of Record for the 32nd America's Cup in 2007. This **Small Column-Wheel Chronograph Lady Team Watch** comprises features including a 5' + 5' indicator serving to count down the crucial last 10 minutes prior to the start of a regatta. Hublot offers its Big Bang model with a smaller yet nonetheless imposing diameter of 41mm. The Aspen version combines a white ceramic case with a matching rubber strap, while the **Porto Cervo** expresses the fusion theme with a red-gold case and red rubber strap, admirably set off by a bezel boasting round or baguette-cut diamonds.

5

## The mono-pusher chronograph

In a more masculine vein, we are observing the return of the mono-pusher chronograph, which requires an extremely sophisticated construction in that the single pusher must successively handle the start, stop and reset functions. Among the gems in this category is the **Villeret Single Push-piece Chronograph** by Blancpain, which comes in versions including a beautiful chocolate-colored version. Jaquet Droz, a brand that specializes in original display modes, offers an extremely elegant variation on this

1

2

great classic by presenting the first mono-pusher chronograph endowed with an hour counter as well as an off-centered hour-minute indication, thus the dial's center is occupied exclusively by the chronograph seconds hand. This **Chronographe Monopoussoir Hommage La Chaux-de-Fonds 1738**, driven by a self-winding column-wheel movement, comes in several versions including ones with stone or enameled dials. In 2006 De Bethune presented a mono-pusher chronograph combining a wealth of performance and innovations, including a patented isochronic balance-spring, a new oscillating system, ultra-light titanium bridges with triple parachute, an original rating system, a double barrel guaranteeing a 172-hour power reserve and more! Meticulous care was devoted to ensuring the readability of the **DB21 Maxichrono's** dial with its five co-axial hands: chronograph functions are read from the outer part of the dial so as not to interfere with the time display, while the seconds hand rotates in 30-second counters to provide a clearer graduation.

Even quartz watches have been enriched with complication mechanisms. Quinting's **Chronographe Mystérieux's** hands are driven by twelve sapphire disks and appear to be floating in mid-air.

3

▸▸ 1 *De Bethune, DB21 Maxichrono*
*The Maxichrono by De Bethune features a dial with co-axial hands as well as several technical innovations in the movement.*

▸▸ 2 *Jaquet Droz, Chronographe Monopoussoir Hommage La Chaux-de-Fonds 1738*
*Jaquet Droz's Chronographe Monopoussoir Hommage La Chaux-de-Fonds 1738.*

▸▸ 3 *Chronographe Mystérieux*
*Quartz goes transparent in the Mysterious Chronograph by Quinting.*

# A few legendary chronograph movements

**• Zenith El Primero**

The famous El Primero movement heralded the era of the self-winding chronograph based on a different principle in 1969. The El Primero is a classic integrated chronograph movement with column-wheel control and ball-bearing-mounted central rotor. For a number of years, it was the only mechanical chronograph movement capable of measuring short times to within 1/10th of a second; in order to achieve this level of precision, Zenith doubled the oscillating frequency of the balance to reach 36,600 vibrations per hour. The El Primero movement has been back in the limelight in recent years, effectively powering Zenith's latest collections. It has been combined with various complications, including a tourbillon, and is also found in chronographs produced by other brands within the LVMH group to which Zenith belongs.

**• Breitling Caliber 11**

Stemming from cooperation between Breitling, Büren and Heuer-Leonidas, the Caliber 11 (according to the Breitling reference) is the other "first" self-winding chronograph movement presented the same year as the El Primero. However, Caliber 11 is based on a very different construction from that of its rival. Its design engineers opted for a modular system in which the independent chronograph mechanism is added to the base movement. This chronograph's "control center" is a cam system that is easier to build and more affordable than a column-wheel mechanism. Caliber 11 is also distinguished by its off-centered micro-rotor. It is the forerunner of most self-winding chronograph movements used today.

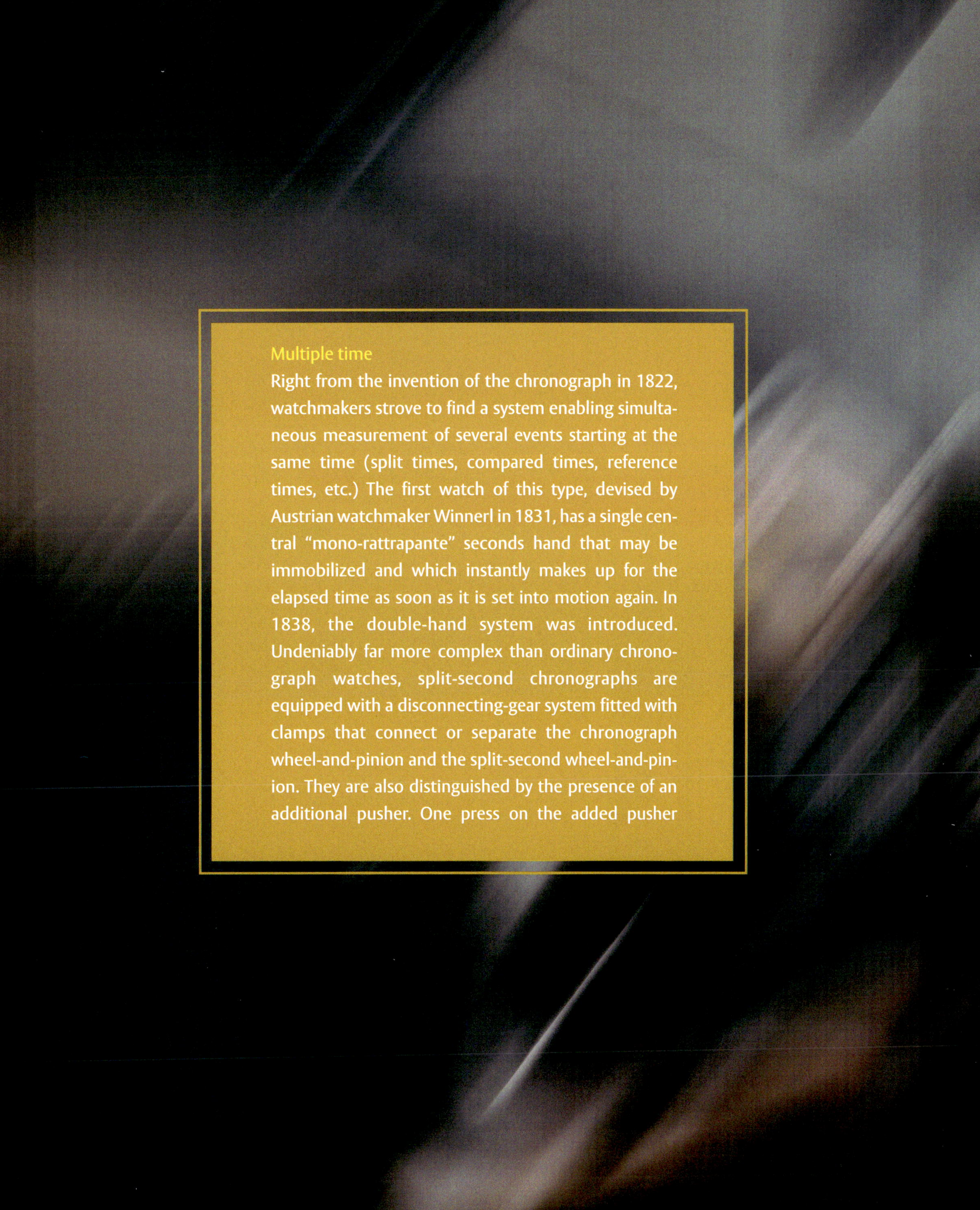

### Multiple time

Right from the invention of the chronograph in 1822, watchmakers strove to find a system enabling simultaneous measurement of several events starting at the same time (split times, compared times, reference times, etc.) The first watch of this type, devised by Austrian watchmaker Winnerl in 1831, has a single central "mono-rattrapante" seconds hand that may be immobilized and which instantly makes up for the elapsed time as soon as it is set into motion again. In 1838, the double-hand system was introduced. Undeniably far more complex than ordinary chronograph watches, split-second chronographs are equipped with a disconnecting-gear system fitted with clamps that connect or separate the chronograph wheel-and-pinion and the split-second wheel-and-pinion. They are also distinguished by the presence of an additional pusher. One press on the added pusher

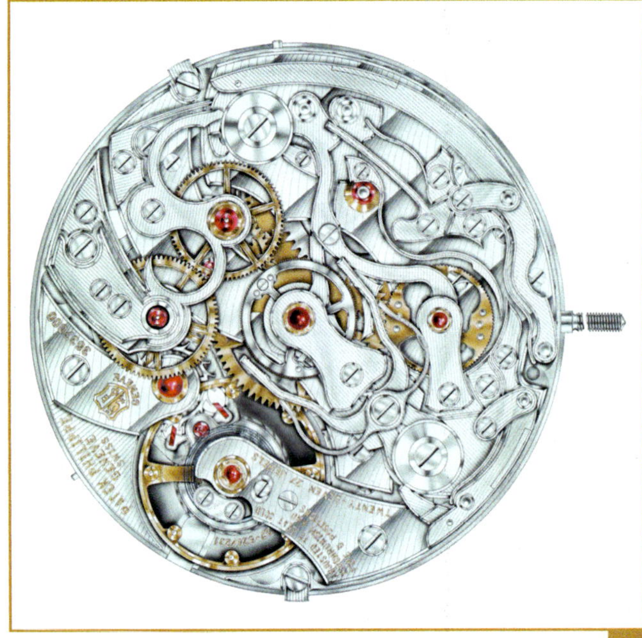

1

stops the split-second hand, and another press on the same pusher sends the temporarily motionless hand rushing to catch up with the first hand, after which the two superimposed seconds hands pursue their rotation together. The operation may be repeated as often as desired. This function proves particularly useful in sports events involving several participants.

### The ultimate chronograph mechanism

Representing an ultra-sophisticated version of the mechanical chronograph, the split-second chronograph is one of the most eloquent demonstrations of horlogical expertise. Prestigious brands are thus keen to present one (or several) models of this type within their collections. In order to fulfill the expectations of an increasingly demanding clientele, watchmakers present an impressive range of innovations and exclusive features—whether in terms of the mechanism itself, the display or the materials. This strong comeback of the split-second chronograph owes much to the current craze for chronographs in general. It also stems from watchmakers' desire to stay off the beaten track and to stand out from the crowd by proving their excellence in fields less "crowded" than those of the tourbillon or the perpetual calendar. Despite this growing success, split-second chronographs—and especially those driven by proprietary movements—are still a relative rarity. They are often combined with other prestigious complications such as the tourbillon.

### Record slimness

In 2005, Patek Philippe reasserted its mastery of horlogical complications by presenting the thinnest split-second column-wheel chronograph movement ever made. This new caliber, dubbed the CH R 27-525 PS and traditionally crafted in limited quantities by the Haute Horlogerie workshops within the manufacture, was also the first wrist-sized chronograph movement

▸▸ 1 *Patek Philippe, Chronographe à Rattrapante Ref. 5959 (nouveauté 2005)*
*In 2005, Patek Philippe presented Ref. 5959, equipped with the thinnest column-wheel split-second chronograph ever made.*

▸▸ 2 *A. Lange & Söhne, Tourbograph Pour le Mérite*
*The Tourbograph Pour le Mérite by A. Lange & Söhne combines a split-second chronograph, a tourbillon and a fusée and chain energy transmission system.*

▸▸ 3 *A. Lange & Söhne, Lange Double Split*
*Lange Double Split by A. Lange & Söhne is the first authentic wrist chronograph with double split-seconds, one for the seconds and the other for the minutes.*

developed and produced entirely by Patek Philippe. Despite its thickness of just 5.25mm, it incorporates all the signature features of the noblest traditions and, like all Patek Philippe mechanical movements, bears the famous Geneva Seal. It is distinguished by its double-pusher system (instead of the three normally used for split-second chronographs) and by several patented characteristics, including the new teeth profiles that considerably enhance the operating performance of the chronograph and reduce wear on the mechanism. Noblesse oblige, Patek Philippe has housed this movement within a platinum case designed in the classic style of officer watches.

## Technical feats

In its Tourbograph Pour le Mérite, produced to mark the firm's 160th anniversary and the 15th anniversary of its reestablishment, the German brand A. Lange & Söhne combined the split-second chronograph with two of the most complex mechanisms ever devised to improve the long-term precision of watch movements: a tourbillon (see Tourbillon chapter) and a fusée and chain transmission system. This extremely complex mechanism (involving no fewer than 600 parts for the chain alone!) guarantees regular transmission of the energy, whatever the state of winding of the mainspring. The Saxony-based brand also innovated in its Lange Double Split by

1

presenting the first genuine double-split wrist-worn chronograph: with one such hand for the seconds and one for the minutes. This makes it possible to compare two timing operations up to 30 minutes. Girard-Perregaux has endowed its Vintage 1945 XXL Foudroyante Split-second Chronograph with jumping seconds, meaning a hand that sweeps round a subdial in one second, facilitating timing to within 1/8th of a second. Blancpain took on several difficult tasks in creating the Single Push-piece Split-second Chronograph within its Villeret collection—a model that is in fact equipped with two pushers—one for the three chronograph functions and the other for the split-second hand). Finally, Omega's De Ville Rattrapante double column-wheel watch is driven by a chronometer-certified self-winding movement equipped with the famous "co-axial escapement" ensuring excellent rating stability.

2

### A daring new breeze

Split-second chronographs tend to have a fairly classic appearance,

▸▸ 1 *Girard-Perregaux, Vintage 1945 XXL Foudroyante Split-second Chronograph Foudroyante*
*Girard-Perregaux has equipped its Vintage 1845 XXL Foudroyante Split-second Chronograph with a seconds hand that makes a complete turn around the counter in one second.*

▸▸ 2 *Omega, De Ville Rattrapante*
*Omega's De Ville Rattrapante has an exclusive "co-axial escapement" system.*

▸▸ 3 *Blancpain, Villeret Single Push-piece Split-second Chronograph*
*Blancpain's Villeret Single Push-button piece Split-second Chronograph.*

3 ▲

1

2

▸▸ 1 *IWC, Double Chronograph Pilot's Watch*
*The Double Chronograph Pilot's Watch by IWC is equipped with a high-tech ceramic case with anti-magnetic soft-iron cage.*

▸▸ 2 *Richard Mille, RM008*
*Richard Mille's RM008 is a highly contemporary architecture inspired by the world of Formula 1.*

▸▸ 3 *Dubey & Schaldenbrand, Spiral-Verso VIP*
*Dubey & Schaldenbrand's Spiral-Verso VIP's mobile index is a simplified form of split-second mechanism invented by Georges Dubey in 1946.*

as befits such valuable models. But they are starting to treat themselves to some daring new ventures in terms of materials and design. In its Double Chronograph Pilot's Watch, IWC has housed this mechanism in a high-tech scratch-resistant ceramic case fitted with an inner antimagnetic soft-iron cage. Far more iconoclastic by nature, Richard Mille has given the split-second chronograph an entirely new face, just as he had done for the tourbillon. Representing the culmination of five years of research and development, the RM008 is powered by a 444-part movement combining a split-second chronograph and a variable inertia tourbillon. The highly contemporary architecture of the movement, inspired by the world of Formula One motor-racing, combines avant-garde materials (for the column wheels, coupling wheels and titanium levers) with innovative technical solutions aimed at reducing energy consumption and avoiding any jitter or bounce of the split-second hand. The watch also features a torque indicator displaying the state of tension within the barrel spring, an effective means of guaranteeing optimal energy at all times.

### Mobile index: a simplified split-second mechanism

No account of split-second chronographs would be complete without a mention of the "mobile index" created in 1946 by Georges Dubey, co-founder of Dubey & Schaldenbrand. In this simplified form of split-second movement, confined to timing operations of less than one minute, the two chronograph seconds hands are linked by a spiral spring that is visible beneath the watch glass. The split-second hand remains motionless as long as one is pressing the pusher; as soon as the pusher is released, the hand joins the other hand under the impetus of the spring. This original system has been revived by Dubey & Schaldenbrand in models such as the Spiral One. To celebrate its 60th anniversary, the brand based in Les Ponts-de-Martel issued a model named Spiral-Verso VIP featuring an elegant dial graced with six blue half-spheres and a 1946 royal blue mobile index. It also comprises a revolutionary detail in that a press of the two pushers at 12:00 releases the case that then turns at will in any direction, revealing the exquisitely finished, entirely hand-engraved movement.

3

# The ABCs of mechanical chronographs

*What is a chronograph?*

A chronograph is a watch equipped with an additional mechanism enabling one to measure and displays elapsed time. It generally has a central direct-drive hand to count off the seconds, while one or two small additional subdials totalize the minutes and the hours. Modern wrist-worn chronographs are often equipped with two pushers: the one at 2:00 serves to start and stop the chronograph, while the other at 4:00 brings the hands back to zero after a measurement.

*How does a chronograph work?*

A chronograph operates by means of a complex system of levers, springs and wheels. To simplify matters, one might say it consists of two main groups of parts: the control system (pushers + column wheel or cam) the coupling system (clutch-lever, clutch-wheel, chronograph driving wheel and chronograph wheel-and-pinion.

The chronograph mechanism is "grafted" onto the base going-train of the watch. When the start pusher is pressed, the chronograph wheel-and-pinion meshes with the seconds wheel and the seconds hand starts. When the chronograph is stopped, the chronograph wheel-and-pinion disengages from the seconds wheel and the seconds hand stops moving. Resetting is performed by means of a system comprising a sprung hammer and a "heart."

*What is a split-second chronograph?*

Split-second chronographs are equipped with an additional seconds hand enabling measurement of intermediate or split times. They are distinguished by the presence of an additional pusher. Pressing this pusher stops the split-second hand and pressing the same pusher again sends the temporarily motionless hand rushing to "catch up" with the first hand, after which the two superimposed hands pursue their rotation together. This highly sophisticated model is one of the most difficult to make, on a par with a tourbillon or a minute repeater.

The mechanism is based on a two-armed clamp that releases or immobilizes the spit-second wheel and pinion, along with a "split-second heart" linked to a spring. When the chronograph

is in operation, the chronograph drives the split-second wheel and pinion and the hands move in perfect synchronization. When the split-second pusher is pressed, the clamps block the split-second wheel-and-pinion, while the chronograph wheel and pinion, to which the split-second heart is fixed, continues to move. When the same pusher is pressed a second time, the clamps release the split-second wheel and pinion and the spring attached to the split-second heart brings the split-second hand back beneath the chronograph hand instantly.

*What is a mono-pusher chronograph?*

Before the invention of the second reset pusher (1934), both pocket and wrist-worn chronographs had just one pusher, generally housed in the crown. This mono-pusher served to execute successively—and always in the same order—the three functions indispensable to using a chronograph: start, stop and reset. This sophisticated construction, also called a three-stroke chronograph, has recently been reinterpreted by several major watch brands in wrist-worn chronographs

*What is the difference between a chronograph and a chronometer?*

The word "chronometer" is often mistakenly used to refer to a chronograph, but they are in fact two very different things. A chronometer is a watch with a movement that has earned an official document certifying its extreme precision. This certificate is awarded by the COSC (Swiss Official Chronometer Testing Institute) after fifteen days of extremely rigorous testing in five positions and at three different temperatures. A chronograph may be chronometer-certified, but not all chronometers (far from it!) are chronographs.

*What is the difference between a column-wheel and a cam type chronograph?*

The difference relates to the "control group" of parts that guides the "start-stop-reset" functions of the chronograph.

The column wheel, featuring one-piece construction, comprises a toothed section called a ratchet and columns placed perpendicular to this toothing. This system of alternating spaces and teeth serves to control the various movements of the levers that rest against a column or are positioned between two columns. Each time a pusher is pressed, the wheel turns one notch to take the appropriate position. The column wheel, while undeniably the noblest version of the mechanical chronograph, is far trickier to make and more expensive than a cam system. However, according to many specialists, it guarantees far smoother and more accurate operation.

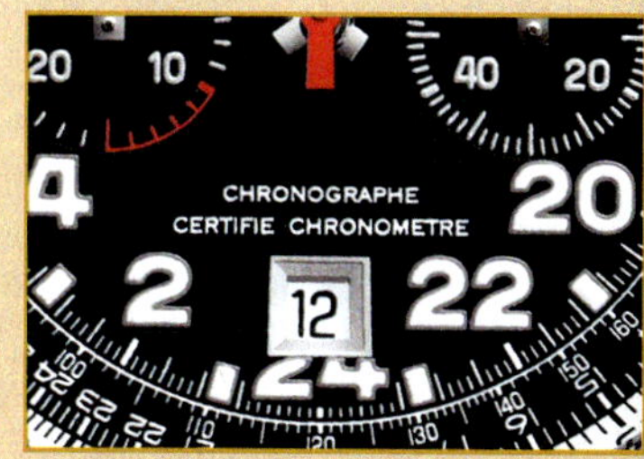


The cam, a much simpler construction, is made up of two parts; the lower "navette," or shuttle, and the upper

shuttle. This geometry makes it possible to control the levers guiding the various chronograph functions. Most current mechanical chronographs (including the ETA Valjoux 7750 movement) are equipped with a cam system, which is easier to make and to adjust, and therefore more economical.

*What is the difference between a modular chronograph and an integrated version?*

In a modular construction, the chronograph mechanism is totally independent from the base movement. It is mounted on an additional plate that is fixed to the dial side of the movement, which means the chronograph mechanism cannot be revealed through a transparent caseback. This solution makes it possible to associate a chronograph mechanism with an existing movement without needing to develop a new caliber. Cam chronographs are generally modular constructions.

In an integrated construction, the chronograph components are placed directly on the bridge side of the movement. This arrangement enables producers to offer transparent casebacks providing an opportunity to admire the chronograph mechanism. Column-wheel movements are integrated constructions.

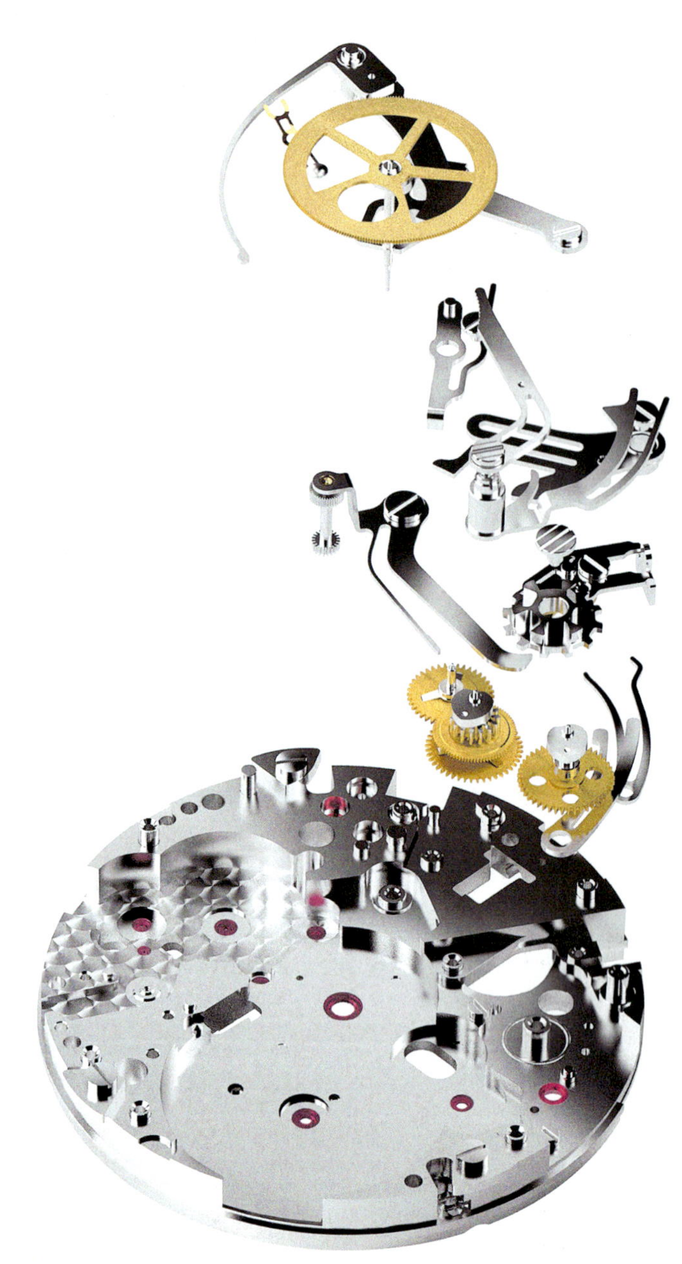

*How precisely can a mechanical chronograph measure times?*

The measurement and display precision of a mechanical chronograph depends on the frequency of the regulating organ, meaning the number of vibrations per hour (one oscillation = two vibrations) of the sprung balance. The higher the frequency, the more the chronograph can measure small fractions of a second.

| Vibrations/hour | Frequency (Hertz) | Smallest measurable time interval |
|---|---|---|
| 18,000 | 2.5 Hz | $1/5^{th}$ of a sec. |
| 21,600 | 3 Hz | $1/6^{th}$ of a sec. |
| 28,800 | 4 Hz | $1/8^{th}$ of a sec. |
| 36,000 | 5 Hz | $1/10^{th}$ of a sec. |
| 360,000 | 50 Hz | $1/100^{th}$ of a sec. |

Watchmakers have constantly striven to optimize the measurement precision of mechanical chronographs. In 1969 with its famous El Primero caliber, Zenith presented the first movement that was accurate to within $1/10^{th}$ of a second; in 2005, TAG Heuer shifted into higher gear by introducing its Chronograph Caliber 360 boasting $1/100^{th}$ of a second precision.

*What is the flyback function?*

The flyback function enables one to instantly launch a new timing operation even when the chronograph is already in operation. All the user need do is directly activate the reset pusher (generally positioned at 4:00) without first needing to stop the hands: the three chronograph hands instantly return to zero and then automatically resume a new measurement. The two pushers may also be used in the same way as on a traditional chronograph: start and stop at 2:00, reset at 4:00. However, in this case, the hands will remain motionless when they are brought back to zero.

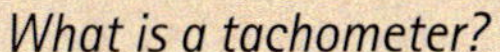

*What is a tachometer?*

A tachometer, or a "tachometric scale," is a graduation on the bezel or the flange which,

when used in conjunction with the chronograph, enables measurement of average speeds in km/h or another unit. For example, a driver presses the chronograph start pusher when passing a milestone; he subsequently presses the same pusher to stop the chronograph at the following milestone. On the graduated scale to which the chronograph seconds hand is pointing, the driver may then read off the average speed clocked over the distance covered.

*What is a foudroyante or "jumping seconds" function?*

The foudroyante or "jumping seconds" function consists of a seconds hand that performs a complete rotation around a subdial in one second—while sometimes making four or five jumps per second. This system serves to measure time to within one-quarter, one-fifth or even one-eighth of a second.

## A. LANGE & SÖHNE

## DATOGRAPH

A benchmark in the construction of superb mechanical chronographs, A. Lange & Söhne's Datograph houses the manually wound Caliber L951.1 chronograph movement with a frequency of 18,000 oscillations per hour and consisting of 90 meticulously hand-finished parts. The watch features a precise jumping minute counter with a stepped pinion, flyback mechanism, column-wheel chronograph and Lange outsized date with rapid date adjustment. The flyback mechanism allows the user to reset the chronograph hands to zero during an ongoing time measurement and to start a new measurement simply by releasing the pushpiece again, thereby eliminating the stop, reset and restart action.

## A. LANGE & SÖHNE

## THE LANGE DOUBLE SPLIT

This model features two rattrapante hands: one for the seconds and one for the minutes to be stopped. The chronograph and rattrapante hands are flyback hands. With the Lange Double Split, comparative lap measurements of up to 30 minutes are possible for the first time. A. Lange & Söhne has filed for a patent for the disengagement mechanism that allows the chronograph sweep seconds hand to continue revolving while the rattrapante sweep seconds hand is stopped. A true technological achievement, the watch is also equipped with a balance wheel (developed in-house) equipped with poising weights instead of inertia screws. The balance spring is not attached to a hairspring stud, but instead is secured by a balance-spring clamp for which a separate patent registration has been filed. The manually wound, 465-part Caliber L001.1 has 40 jewels.

## A. LANGE & SÖHNE

## THE LANGE DOUBLE SPLIT MOVEMENT

The A. Lange & Söhne Caliber L001.1 manually wound movement is precision adjusted in five positions and features plates and bridges made from untreated cross-laminated German silver. It houses 465 parts and 40 jewels, and is the world's first flyback chronograph with double rattrapante, controlled by classic column wheels. Other functions include jumping chronograph minute counter and rattrapante minute counter, flyback function, disengagement mechanism, hours, minutes, small seconds with stop seconds, power-reserve indicator, tachometer scale, and lap-time measurements between 1/6th of a second and 30 minutes.

## AUDEMARS PIGUET

## LADY ROYAL OAK OFFSHORE ALINGHI - REF. 26076SK.ZZ.D010CA.01

For the many women among the ranks of sailing enthusiasts, this manufacture in Le Brassus has created the Royal Oak Offshore Lady Alinghi chronograph. The large-sized steel case and generous clear-cut lines exude a distinctive character enhanced by 32 diamonds set around the bezel in a sparkling line punctuated by 8 polished-steel hexagonal screws. Three chronograph counters stand out against this dazzling background: the minute counter at 3:00; small seconds at 6:00; and an hour counter at 9:00 adorned with the spiraling Team Alinghi logo in bright red. This hue is echoed by the numerals on the minute circle and accentuates the fact that the Royal Oak Offshore Lady Alinghi is part of a great sporting family. Finally, in an ultimate touch inspired by the prestigious America's Cup competition, the Royal Oak Offshore Lady Alinghi's caseback is engraved with two sails billowing in the wind and the words "Limited Edition." Supplied with gemstone certification.

## AUDEMARS PIGUET

## LADY ROYAL OAK OFFSHORE CHRONOGRAPH - REF. 26048SK.ZZ.D035CA.01

This stainless steel Lady Royal Oak Offshore Chronograph is topped with a bezel set with 32 brilliant-cut diamonds (1.25 carats) and clad in pistachio-green rubber on a matching strap. The timepiece features hours, minutes, date and chronograph. The Royal Oak Offshore logo is engraved on the caseback and its pistachio-green dial features a Royal Oak Mega Tapisserie guilloché pattern with 8 polished cabochon hour markers. Supplied with gemstone certification.

## AUDEMARS PIGUET

## LADY ROYAL OAK OFFSHORE CHRONOGRAPH - REF. 26092OK.ZZ.D080CA.01

This 18K pink-gold Lady Royal Oak Offshore Chronograph in chocolate brown, with bezel, case and lugs set with brilliant-cut diamonds (5.9 carats), features hours, minutes, sweep seconds hand, 30-minute and 12-hour counters, date and chronograph. The Royal Oak Offshore logo is engraved on the caseback and its dial features a Royal Oak Mega Tapisserie guilloché pattern, with 8 diamond hour-markers. Supplied with gemstone certification.

## AUDEMARS PIGUET

## ROYAL OAK CHRONOGRAPH - REF. 25960OR.00.1185OR.01

This Royal Oak Chronograph houses the self-winding Caliber 2385 with rhodium-plated 18K gold rotor, Côtes de Genève decorative pattern and circular graining. Protected by a sapphire crystal, the dial is made with an exclusive, traditional, engine-turned Grande Tapisserie decorative pattern and displays the hours, minutes, date, and chronograph functions.

## AUDEMARS PIGUET

## ROYAL OAK CHRONOGRAPH - REF. 26068BA.ZZ.D088CR.01

Cased in diamond-pavé yellow gold, this Royal Oak Chronograph is set with a total of 449 brilliant-cut diamonds (4.04 carats). Its self-winding Caliber 2385 features a rhodium-plated 18K gold rotor, Côtes de Genève decorative pattern and circular graining. The diamond-pavé dial displays hours, minutes, date, cognac sapphire indexes, mother-of-pearl chronograph counters, and openworked gold hands under a sapphire crystal. Supplied with gemstone certification.

## AUDEMARS PIGUET

## ROYAL OAK OFFSHORE CHRONOGRAPH - REF. 26020ST.00.D091CR.01

Crafted in stainless steel, this Royal Oak Offshore Chronograph displays hours, minutes, chronograph and date. The dial, with the exclusive Extra Grande Tapisserie pattern, bears luminescent hour-markers and hands. The Royal Oak Offshore Chronograph is mounted on a "Hornback" crocodile leather strap with the AP folding clasp in polished steel.

## AUDEMARS PIGUET

## ROYAL OAK OFFSHORE CHRONOGRAPH - REF. 26067BC.ZZ.D002CR.01

Bearing 341 brilliant-cut diamonds (5.86 carats), this white-gold Royal Oak Offshore Chronograph houses the self-winding Caliber 2840. The movement features a rhodium-plated 21K gold rotor, Côtes de Genève decorative pattern and circular graining. The diamond-pavé dial displays hours, minutes, date, satin-brushed chronograph counters and luminescent gold hands. The diamond-pavé case is fitted with a sapphire crystal. Supplied with gemstone certification.

## BOVET

## FLEURIER MANUALLY WOUND SINGLE-BUTTON DOCTOR'S CHRONOGRAPH

BOVET's first chronograph in a 42mm Fleurier case returns to the style of the earliest wrist-chronographs of the late 1920s—a single-button control with the elapsed chronograph minutes and the running small seconds vertically aligned. Shown here in 18K white gold, this type of chronograph with a pulsometer scale is known as a doctor's chronograph. The physician times 15 pulse-beats and reads off the pulse rate per minute on the scale. The chronograph stop, start and zero functions are controlled by the application of successive pressures on a cabochon-topped button coaxial with the crown. The original chronograph system is much easier to use than the more modern two-button chronograph. This model is also available in Ø 39mm and in 18K rose gold.

## BREGUET

## TYPE XXI - REF. 3810BR/92/9ZU

This Type XXI fly-back chronograph is crafted in 18K rose gold and its features include an elapsed-minutes register by a central hand, a self-winding movement with date and subdial for the seconds, a day/night indicator and 12-hour totalizer, while its dial is enhanced with luminous hands and hour-markers. Its case has a graduated turning bezel and screw-locked crown, enabling a water resistance to 100 meters.

## BREGUET

## CLASSIQUE - REF. 5238BB/10/9V6 DD00

This Classique openworked chronograph is crafted in 18K white gold and houses a hand-wound movement with small seconds and 30-minute totalizer. The case features a bezel, lugs and caseband paved with 96 baguettes (approximately 13 carats).

## BREGUET

## MARINE - REF. 5827BB/12/9Z8

This Marine chronograph is crafted in 18K white or yellow gold and houses a self-winding movement with date and small seconds. The silvered gold dial is hand-engraved on a rose-engine, has a 15-minute sector and features central chronograph minutes and seconds. The case features a sapphire caseback and a screw-locked crown, enabling water resistance to 100 meters. The model is available with bracelet as reference 5827BB/12/BZ0 or 5827BA/12/AZ0.

## BREGUET

## CLASSIQUE - REF. 5947BA/12/9V6

This Classique split-seconds chronograph is crafted in 18K yellow or white gold and houses a manually wound movement with small seconds, compensating balance-spring with Breguet overcoil and 48-hour power reserve. Its gold dial, engine-turned and silvered, features a 30-minute elapsed-time counter at 3:00. Its case is enhanced with a sapphire caseback and is water resistant to 30 meters.

## BVLGARI

## DIAGONO PROFESSIONAL CHRONO FLY- BACK - REF. GMT40C6SVD/FB

The Diagonal Professional Chrono Fly-back features an automatic-winding mechanical movement and is a COSC-certified chronometer, finished and decorated by hand, with a 42-hour power reserve, 53 jewels and a vibration rate of 28,800 vph (4Hz). It offers the functions of hours, minutes, small seconds, date, flyback chronograph with hours, minutes, seconds and simultaneous display of 3 time zones. The 40mm brushed stainless steel case is fitted with an antireflective scratch-resistant sapphire crystal, a bi-directional rotating bezel with 24-hour scale, a security screw-down stainless steel crown with case protection, 3 screw-down pushbuttons for chronograph functions and GMT correction, and is water resistant to 100 meters. The dial options are white or silver with a black and yellow flange, 3 black counters for chronograph functions, hand-applied luminescent hour indexes and numeral 12, and black luminescent hands and a black GMT hand with red arrow. The display includes GMT 3 time zones and date at 4:30. The black rubber strap is enhanced with brushed stainless steel links and a triple-fold-over buckle.

## BVLGARI

## DIAGONO PROFESSIONAL CHRONO RATTRAPPANTE - REF. CH40C6GLTARA

The Diagono Professional Chrono Rattrappante features an automatic-winding mechanical COSC-certified hand-decorated movement, with a 42-hour power reserve, 31 jewels and a vibration rate of 28,800 vph. Its functions include hours, minutes, small seconds, and split-second chronograph with hours, minutes and seconds. The 40mm brushed 18K yellow or white-gold case is fitted with an antireflective, scratch-resistant, sapphire crystal, a bezel with engraved tachometric scale, a security screw-down gold crown with case protection, gold pushbutton, a snap-on back, displaying the movement through a sapphire crystal, and is water resistant to 100 meters. The silver dial with vertical treatment displays 3 chronograph counters with gold-plated outline, hand-applied, gold-plated indexes with luminescent dots and luminescent, gold-plated hands, as well as a center second and split-second indication on the flange. The hand-sewn brown or black alligator straps are fitted with an 18K yellow- or white-gold triple-fold-over buckle.

## BVLGARI

## DIAGONO PROFESSIONAL REGATTA - REF. SD40BSV/RE

The Diagono Professional Regatta features a mechanical, automatic-winding 30mm movement with a 42-hour power reserve, 39 jewels and a vibration rate of 28,800 vph, is finished and decorated by hand. Its functions include hours, minutes, fly-back chronograph, minute amplification display and Regatta tactical functions. Its brushed 40mm, stainless steel case is fitted with an antireflective, scratch-resistant, sapphire crystal, a bi-directional rotating bezel with compass scale, a security screw-down stainless steel crown with case protection, two screw-down pushbuttons for chronograph functions, and is water resistant to 300 meters. Its black or white dial with sunburst effect is enhanced with hand-applied luminescent round hour indexes and special minute amplification display at 12 (5 portholes), contrasting with rhodium-plated luminescent hands and a red seconds hand. Special Regatta indicators assist with tactical decisions and are embellished with a minute amplification display. The black rubber strap has brushed stainless steel links and a triple-fold-over buckle.

## CHOPARD

## L.U.C CHRONO ONE CHRONOGRAPH

To mark the 10th anniversary of its creation, Chopard Manufacture wished to demonstrate its perfect mastery of haute horology by creating the L.U.C Chrono One chronograph. The result is a self-winding double-pusher chronograph, with two pushers handling chronograph start, stop and reset, as well as flyback function and small seconds reset on demand. The new L.U.C 10 CF caliber offers 60 hours of power reserve. This exceptional movement is housed in an 18K white-gold case and limited to a numbered series of 100.

## CHOPARD

## MILLE MIGLIA JACKY ICKX 6/24 EDITION 4 - REF. 16/8998

The Mille Miglia Jacky Ickx 6/24 Edition 4 is equipped with an imposing steel case with screw-lock pushbuttons and crowns. This model features a large date display at 12:00. The hour, minute and seconds hands are in rhodium-plated metal with SuperLumiNova for nighttime readability. A tachometric scale is engraved on the bezel (from 60 to 240 km/h). In tribute to Jacky Ickx's six wins in the ultimate endurance Le Mans challenge, the chronograph movement, duly chronometer-certified by the COSC, is distinguished by its 24-hour totalizer and the caseback carrying the symbolic 6/24 engraving. This model is water resistant up to 50 meters. Limited edition of 1,000 pieces in steel.

## CHOPARD

## MILLE MIGLIA SPLIT SECOND - REF. 16/1261

Crafted in 18K rose gold, the generously sized case (Ø 44mm) features resolutely sporting lines. The black dial shows four counters arranged in an extremely dynamic way, with the date at 3:00 oversized and raised to facilitate reading the pointer-type display. Records are measured at the flick of a finger thanks to the pushbutton at 2:00 and the red arrow-type chronograph seconds hand. The pushbutton at 8:00 serves to stop the split-second hand to measure a split time, and then make it "catch up with" the chronograph seconds hand to continue the timing operation. Self-winding movement, chronometer-certified by the COSC. This model is water resistant up to 50 meters. Limited edition of 250 timepieces in rose gold.

## CHOPARD

## SQUARE CHRONOGRAPH, REF. 16/8961

Chopard's new steel chronograph is powered by a self-winding COSC-certified movement endowed with approximately 40 hours of power reserve and beating at 28,800 vibrations per hour. The black dial is framed by a bezel engraved with a tachometric scale for reading off speeds of 55 to 400 km/h. This 39mm x 46.5mm model is equipped with a 60-second counter at 3:00, date at 4:30, a 12-hour counter at 6:00, a 30-minute counter at 9:00, and a 1/4th of a second chronograph hand. The glare-resistant sapphire crystal ensures perfect readability of the dial featuring luminescent dots. On a black alligator leather strap with folding clasp, this model is water resistant to 50 meters.

## DANIEL ROTH

## CHRONOMAX - REF. 347.Y.70.763.CM.BD

The Chronomax features an entirely hand-decorated automatic chronograph manufacture movement with minute and hour counters, small seconds hand and date at 6:00, Optimax® indicator at 12:00, offering a 45-hour power reserve. It is housed in a curved white-gold case (44x41mm), enhanced by a white curved dial presenting counter-rings and small seconds on a ruthenium plate. The unique feature of this chronograph timepiece is the indication of the measurement of barrel spring torque, which shows the optimal chronograph conditions of use (Optimax®). This model is water resistant to 3atm.

## DANIEL ROTH

## PERPETUAL CALENDAR CHRONOGRAPH MOON PHASE - REF. 379.Y.50.193.CC.BD

The Perpetual Calendar Chronograph Moon Phase features a self-winding movement, based on a Frédéric Piguet 1185 column-wheel chronograph movement, offering a 40-hour power reserve. This movement is entirely decorated with high-end finishing and features an 18K gold guilloché rotor. Housed in a 5N red-gold curved case, this model features chronograph, perpetual calendar and moonphase functions on a Clous de Paris guilloché dial with circular-guilloché counters. It is water resistant to 3atm.

## EBEL

### 1911 BTR AUTOMATIC CHRONOGRAPH

Twenty years after introducing its emblematic 1911 collection, the Architects of Time present the ultimate Ebel series for men: a highly contemporary, entirely mechanical and eminently masculine collection of timepieces, the 1911 BTR. Featured here, the Ebel 1911 BTR Automatic Chronograph sports a bold, 44.5mm stainless steel case with rubber features, along with vivid red and black accents, complemented by a supple, hand-sewn alligator strap with red top-stitching that offers a supremely comfortable fit and feel. It is powered by the COSC-certified, proprietary Ebel Automatic Chronograph Caliber 137 movement.

## GERALD GENTA

## ARENA CHRONO QUATTRO RETRO - REF. ABC.Y.80.290.CN.BD

The Arena Chrono Quattro Retro features a self-winding manufactured chronograph movement offering a 45-hour power reserve and including a world premiere combination of a jumping-hour mechanism and four retrograde functions. The multi-layer perforated dial features a jumping hour, retrograde minutes and date, as well as two retrograde chronograph functions: hours and minutes. The movement is housed in a new titanium sport case with fluted caseband, screw-down beaded crown and two profiled chronograph pushpieces. This model is water resistant to 10atm.

## GERALD GENTA

## ARENA CHRONO QUATTRO RETRO GOLD - REF. ABC.Y.66.295.CN.BD

The Arena Chrono Quattro Retro Gold, features an exclusive self-winding chronograph movement entirely decorated by hand, with old-gold finish, It also offers a jumping hour, retrograde minutes and date, as well as a chronograph function with retrograde hour-counter at 9:00, retrograde minute-counter at 3:00 and large central seconds hand. It is housed in a polished white-gold case with fluted case middle, fluted gold pushbuttons and brushed-finish tantalum bezel, enhanced by a multi-level dial with satin-brushed counters, matte with holes for the central zone, and applied numerals. It is water resistant to 10atm.

## GERALD GENTA

## OCTO CHRONO QUATTRO RETRO GOLD - REF. OQC.Z.50.581.CN.BD

The Octo Chrono Quattro Retro features an exclusive self-winding GG7800 chronograph movement, entirely decorated by hand, old-gold finish. It offers a jumping hour, retrograde minutes and date; chronograph function with retrograde hour counter at 9:00, retrograde minute counter at 3:00 and large central chronograph seconds hand. It is housed in an octagonal satin-finished 5N red-gold or white-gold case, polished streamlined pushbuttons and round bezel with circular satin-brushed finish on its upper part and polished octagonal base. Multi-level dial with white-lacquered chronograph counters, black- and white-lacquered minute and date displays and vertical satin-brushed central zone. It is water resistant to 10atm.

## GIRARD-PERREGAUX

## CHRONOGRAPHE SPORT MONTE CARLO 1976 - REF. 49540

Released in a limited edition of 1,000 pieces, this Chronographe Sport Monte Carlo 1976 houses the mechanical automatic-winding GP 019C0 caliber with 48 hours of power reserve that beats at 28,800 vibrations per hour and holds 62 jewels. The movement is decorated with a Côtes de Genève pattern and is realized and finished by hand in-house. Its three-piece, 40mm stainless steel case bears an antireflective curved sapphire crystal, screw-down crown, and back fastened by seven screws featuring an engraving dedicated to the Munari-Maiga victory at the Monte Carlo Rally 1976.

## GIRARD-PERREGAUX

## CLASSIQUE ELEGANCE CHRONO WW.TC - REF. 49805

This Classique Elegance Chrono ww.tc houses the mechanical automatic-winding GP 33C0 caliber with 63 jewels. Beating at 28,800 vibrations per hour, it offers 46 hours of power reserve and is decorated with a Côtes de Genève pattern. Built and finished in-house, this ww.tc offers hour, minutes, small seconds, date, world time, 24-hour indication and 3-counter chronograph on an anthracite dial. Its three-piece, 43mm 18K white-gold case is fitted with an antireflective curved sapphire crystal, screw-down crowns, exhibition sapphire crystal caseback and is water resistant to 3atm.

## GIRARD-PERREGAUX

## PETIT CHRONOGRAPH - REF 80440D11AB11-BKBA

Powered by the mechanical automatic-winding GP 030C0 caliber (base and chronograph module with column wheel), this ladies' Petit Chronograph offers 36 hours of power reserve. The 38-jeweled movement beats at 28,800 vibrations per hour and is realized, mounted, adjusted, finished and decorated entirely by hand in-house. In addition to the hour, minutes and small seconds, it offers a 2-counter chronograph. Water resistant to 3atm, the 32mm watch is cased in steel and topped with a bezel set with 48 diamonds weighing approximately 0.96 carat. The caseback is fastened by five screws and displays the movement through a sapphire crystal.

## GIRARD-PERREGAUX

## SPORT CLASSIQUE LAUREATO EVO$^3$ CHRONOGRAPH - REF. 80180

Housing the mechanical automatic-winding GP 033C0 caliber (GP 3300 base and DD 2070 module), this Sport Classique Laureato Evo$^3$ Chronograph offers 46 hours of power reserve, 52 jewels, beats at 28,800 vibrations per hour, and is beveled and decorated with a Côtes de Genève pattern. Completely built by hand in-house, the watch offers small second, date, dual time, 24-hour indication and 3-counter chronograph, tachymeter scale and luminescent hands under a curved sapphire crystal. Water resistant to 10atm, the three-piece, 44mm stainless steel case is fitted with a screw-down crown and a sapphire crystal caseback through which the movement is visible.

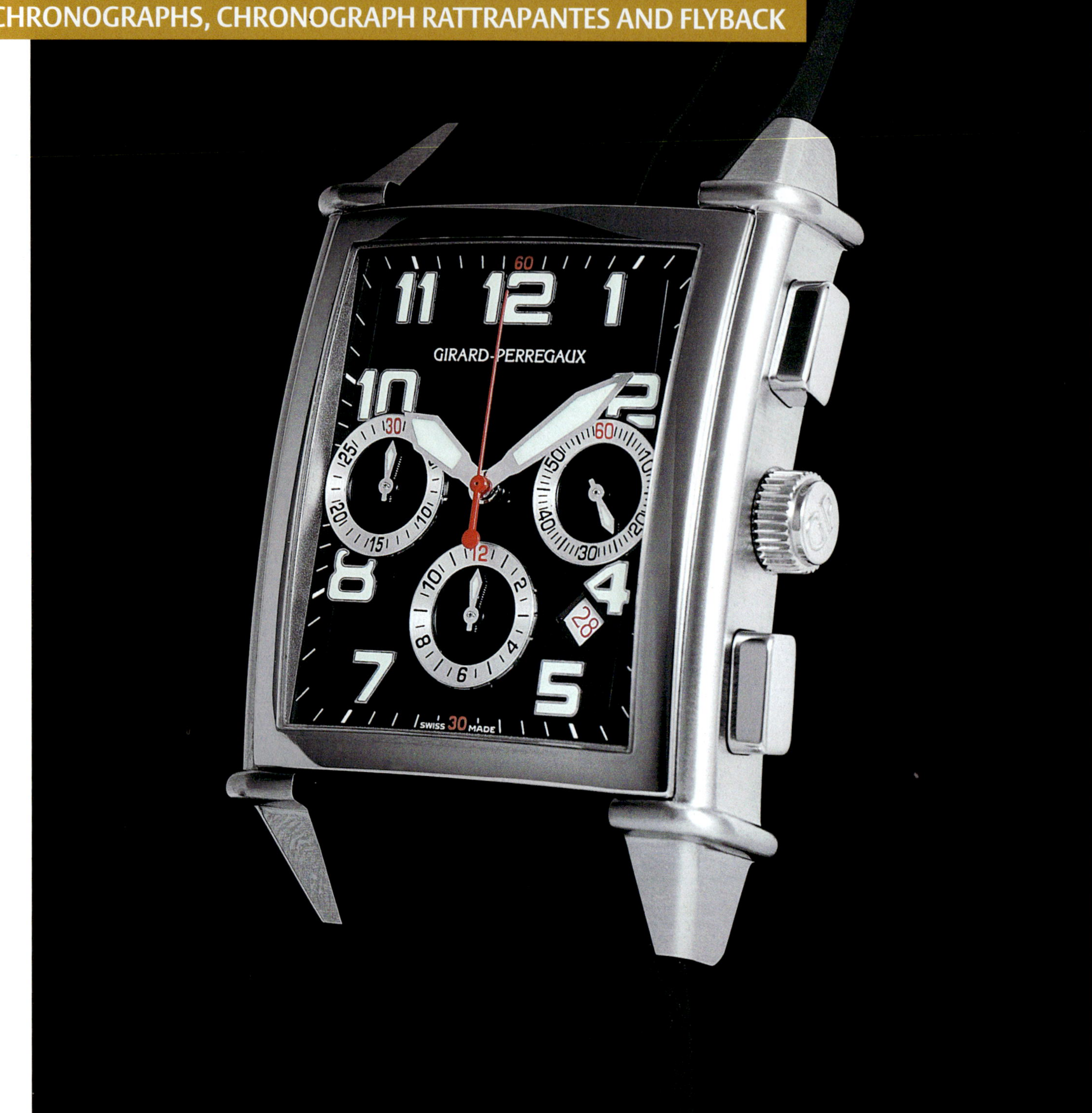


## GIRARD-PERREGAUX

## VINTAGE 1945 XXL CHRONOGRAPH - REF. 25840

Housing the mechanical automatic-winding GP 033C0 caliber with approximately 46 hours of power reserve, the 63-jeweled Vintage 1945 XXL Chronograph offers hour, minutes, small seconds, date and 3-counter chronograph. Beating at 28,800 vibrations per hour, the movement is beveled and decorated with a Côtes de Genève pattern entirely by hand in-house. The watch is encased in steel in a brushed and polished finish with an antireflective sapphire crystal and a sapphire crystal caseback. The semi-matte black dial features luminescent hands and numerals. Released in a 1,000-piece limited edition, the watch is water resistant to 3atm.

## GIRARD-PERREGAUX

## VINTAGE 1945 KING SIZE GMT CHRONOGRAPH - REF. 29975

This elegant 18K white-gold Vintage 1945 King Size GMT Chronograph is anatomically curved to fit the wrist. This multi-functioned watch indicates hour, minute, small second at 3:00, date, dual-time, 24-hour readout at 12:00, and a 3-counter chronograph. The white minute track has a red scale with 5-minute graduation. Visible through a sapphire crystal caseback, the 61-jeweled mechanical automatic-winding GP 033CO caliber has 46-hour power reserve, 28,800 vph, and gold rotor and is beveled and decorated with Côtes de Genève.

## GUY ELLIA

## JUMBO CHRONO

The first round men's watch created by GUY ELLIA, the Jumbo Chrono is based on an interesting technological approach demonstrating that it is possible to combine aesthetic balance with masculine detail and a fine mechanical movement. The Jumbo Chrono bares an exceptional Ø 50mm size that is rarely seen in the world of luxury watches. The Jumbo Chrono's slate-gray oscillating weight is unique and the impressive discovery dial reflects a strong graphic design: chronograph seconds hand is in center, hour and minute are at 12:00, date at 2:00, 30-minute counter at 3:00, seconds at 6:00, and 12-hour counter at 9:00. This model is created in white gold.

## GUY ELLIA

## JUMBO CHRONO

The first round men's watch created by GUY ELLIA, the Jumbo Chrono is based on an interesting technological approach demonstrating that it is possible to combine aesthetic balance with masculine detail and a fine mechanical movement. At Ø 50mm, the Jumbo Chrono bares an exceptional size that is rarely seen in the world of luxury watches. The Jumbo Chrono's slate-gray oscillating weight is unique and the impressive discovery dial reflects a strong graphic design: chronograph seconds hand is in center, hour and minute are at 12:00, date at 2:00, 30-minute counter at 3:00, seconds at 6:00, and 12-hour counter at 9:00. Shown in yellow gold.

## GUY ELLIA

## JUMBO CHRONO

The first round men's watch created by GUY ELLIA, the Jumbo Chrono (shown here in black gold) is based on an interesting technological approach demonstrating that it is possible to combine aesthetic balance with masculine detail and a fine mechanical movement. With a 50mm diameter, the Jumbo Chrono bares an exceptional size that is rarely seen in the world of luxury watches. The Jumbo Chrono's slate-gray oscillating weight is unique and the impressive discovery dial reflects a strong graphic design: chronograph seconds hand is in center, hour and minute are at 12:00, date at 2:00, 30-minute counter at 3:00, seconds at 6:00, and 12-hour counter at 9:00.

## HUBLOT

## BIGGER BANG - REF. 308.TX.130.RX.094

Crafted in platinum PT950, this Bigger Bang's skeleton dial is framed by a bezel holding 48 white baguette diamonds and a round-diamond-pavé case. Total diamond weight is 3.9 carats and this version is limited to just 3 pieces. It is powered by the new chronograph / flying tourbillon HUB 1400 CT movement, visible through a sapphire crystal caseback.

## HUBLOT

## BIGGER BANG - REF. 308.PM.130.RX

Combining 18K red gold and black ceramic, this model was released in an 18-piece limited editon. The dial reveals the HUB 1400 CT's chronograph mechanism and column wheel at 12:00 and is protected by a sapphire crystal with double-sided antireflective treatment. A special 3-piece limited edition is available with a bezel set with 48 white baguette diamonds and a round-diamond-pavé 18K red-gold case (total diamond weight: 3.9 carats).

## IWC

## AQUATIMER CHRONO-AUTOMATIC - REF. IW371933

This watch features a mechanical chronograph movement and a mechanical rotating inner bezel. It houses an IWC 79320 caliber with 25 jewels and offers a day and date indicator and small seconds hand with stop function. By incorporating IWC's extensive experience in the technically demanding realm of pressure resistance, this watch is water resistant up to 120 meters.

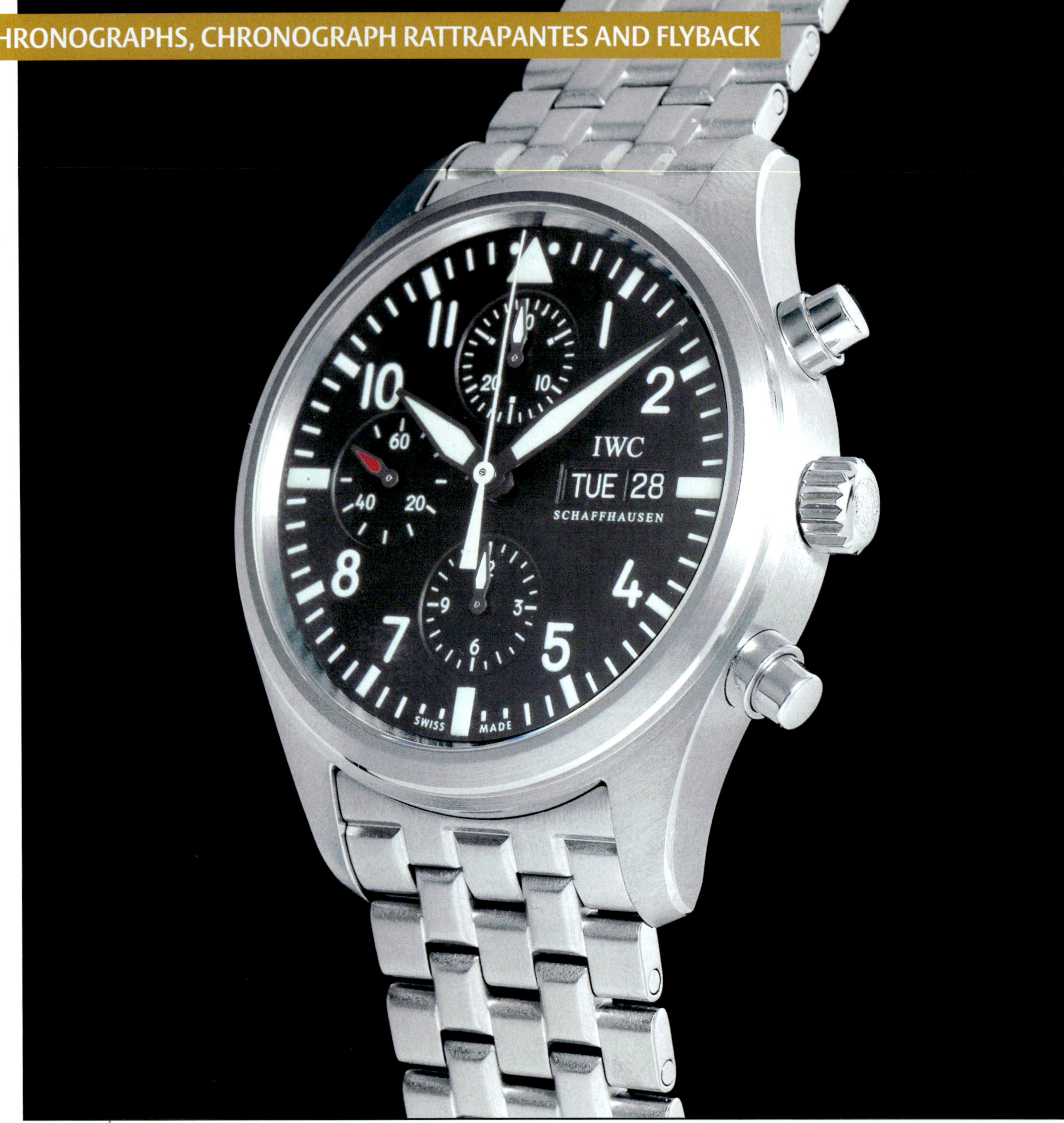


## IWC

## PILOT'S WATCH CHRONO-AUTOMATIC - REF. IW371701

The Pilot's Watch Chrono-Automatic features a robust mechanical chronograph with day and date indicator. It houses the IWC 79320 caliber with 25 jewels, 44 hours of power reserve and features a soft inner iron case to protect against magnetic fields. With its cockpit look, this watch has outstanding readability.

## IWC

## PILOT'S WATCH DOUBLE CHRONOGRAPH RATTRAPANTES EDITION TOP GUN REF. IW3799

This limited edition features a mechanical chronograph movement and houses an IWC 79230 caliber with 29 jewels. It features a softer inner iron case to protect against magnetic fields. The black dial with recessed subdials has contrasting white numerals with stop-second hands in signal-red to enhance readability.

## IWC

## PORTUGUESE CHRONO- AUTOMATIC - REF. IW371431

This 18K white-gold Portuguese Chrono-Automatic features the classic pocket watch format on the wrist. It features a mechanical chronograph movement and a small seconds hand with stop function. This watch houses the IWC 79350 caliber with 31 jewels and 44 hours of power reserve.

## IWC

## SPITFIRE CHRONO-AUTOMATIC - REF. IW371702

Description: A tribute to the legendary Spitfire aircraft of the 1930s, this watch features a robust mechanical chronograph. It offers a day and date indicator and houses an IWC 79320 caliber with 25 jewels and a 44-hour power reserve. This watch also features a soft inner iron case to protect against magnetic fields and drop in air pressure.

## JAEGER-LECOULTRE

## AMVOX2 CHRONOGRAPH CONCEPT

One of Jaeger-LeCoultre's most dramatic innovations, the Amvox2 Chronograph Concept driver's watch features a patented technical breakthrough: the vertical-trigger chronograph. This mechanism makes it possible to start, stop and reset the chronograph simply by pressing on the sapphire crystal, eliminating the need for pushbuttons. Available in a limited series, the titanium and steel watch houses the recently developed chronograph Caliber 751B. Starting is achieved by pressing the crystal at 12:00, reset at 6:00. The watch consists of a ball-joint system that enables the case and bezel to pivot away from the shoulders of the watch, thereby activating a series of levers that transmit impulses to control the chronograph. There is also a three-position cursor that locks the column-wheel chronograph so it is not triggered by the wearer's normal arm motions. The 41-jeweled movement consists of 272 parts.

## JAEGER-LECOULTRE
## MASTER COMPRESSOR CHRONOGRAPH

The Master Compressor Chronograph houses the automatic Jaeger-LeCoultre Caliber 751 with 272 parts and 41 jewels. Beating at 28,800 vibrations per hour, it offers 72 hours of power reserve. The chronograph features an hour and minute counter and has a center second readout and offers date, and a tachometric scale on the inner bezel; the case features the 1000 Hours Control seal on the caseback.

## JAEGER-LECOULTRE

## MASTER COMPRESSOR EXTREME WORLD CHRONOGRAPH

This Master Compressor Extreme World Chronograph houses the mechanical automatic Jaeger-LeCoultre Caliber 752, which is crafted, assembled and decorated by hand. It beats at 28,800 vibrations per hour, offers 72 hours of power reserve and consists of 279 parts and 41 jewels. The chronograph indicates hours, minutes, date, running indication and simultaneous indication of the 24 time zone. The dial is black for the titanium-and-steel version, gray for the 200 titanium-and-platinum pieces. The crown, equipped with a compression key, starts the watch and adjusts the date. This Master Compressor is water resistant to 100 meters, equipped with a shock absorber and features the 1000 Hours Control seal on the caseback.

## JAEGER-LECOULTRE

## REVERSO SQUADRA CHRONOGRAPH GMT

The 18K pink-gold swivel case of the Reverso Squadra Chronograph GMT houses the automatic Manufacture Jaeger-LeCoultre Caliber 754 beating at 28,800 vibrations per hour and endowed with a 65-hour power reserve. Its guilloché silvered dial shows luminescent hours and minutes, an exclusive large contrasting date at 12:00, a 30-minute counter at 3:00 and an hour counter at 9:00, as well as a GMT dual time-zone display at 6:00 that may be adjusted at any time using the crown. The sapphire crystal caseback reveals the details of the movement decorated featuring vertical Côtes de Genève, blued screws and an engraved, openworked oscillating weight. Water resistant to 50 meters, the Reverso Squadra Chronograph GMT is fitted with a pink-gold bracelet.

## JAEGER-LECOULTRE

## REVERSO SQUADRA WORLD CHRONOGRAPH

In this contemporary titanium version, the Reverso Squadra World Chronograph is powered by the new Jaeger-LeCoultre Caliber 753, which beats at 28,800 vibrations per hour and is endowed with a 65-hour power reserve. On the front, a guilloché black dial with white transferred numerals displays the hours and minutes, the chronograph functions with hour counter at 12:00 as well as small seconds at 6:00. The other side of the Reverso Squadra World Chronograph is distinguished by a central disk, enabling instant read-off of the time around the world on a 24-hour scale, with the time zones represented by city names appearing on the black dial. The Reverso Squadra World Chronograph is worn on an alligator leather strap with triple-blade folding clasp.

## JAQUET DROZ

## CHRONO MONOPOUSSOIR ÉMAIL NOIR ABSOLU - REF. J007634209

The Chrono Monopoussoir is the world's first single pushbutton chronograph to feature an hour counter and hour-minute indication that are both off-centered. Supported by a powerful dial design on which only the chronograph seconds hand runs from the center, the column-wheel Chrono Monopoussoir features functions that are all controlled by a single pushbutton set into the crown.

## MAURICE LACROIX

## MASTERPIECE LE CHRONOGRAPHE

Maurice Lacroix presented its first own manufacture movement—the ML 106 caliber in the Masterpiece Le Chronographe. This heralds the start of a new era for the Swiss brand. The chronograph movement is unique; its majestic proportions of Ø 36.6mm and the technology behind it are captivating. This technology is a marriage between traditional elements, including a quality column-wheel and an innovative, specially developed lever mechanism for stopping and zeroing, for which a patent is pending. The first 250-piece limited edition of the Masterpiece Le Chronographe is released in pink-gold case.

## PARMIGIANI FLEURIER

## KALPAGRAPH - REF. PF 003912.01

Shown in polished steel on a Hermès calfskin strap, the new 53.4x39.2mm Kalpagraph is powered by the automatic-winding PF Caliber 334.01 with 50-hour power reserve, 68 jewels, and beating at 28,800 vph. (The gold versions have a 22K gold oscillating weight.) The snailed chronograph counters are at 3:00 (small seconds), 6:00 (12-hour) and 9:00 (30-minute). The dial is decorated with "velouté" and "satin sun" finishes, applied indexes and luminescent delta-shaped hands. Each sapphire crystal caseback is individually numbered. Versions are available on rubber straps and in rose gold or 950 palladium.

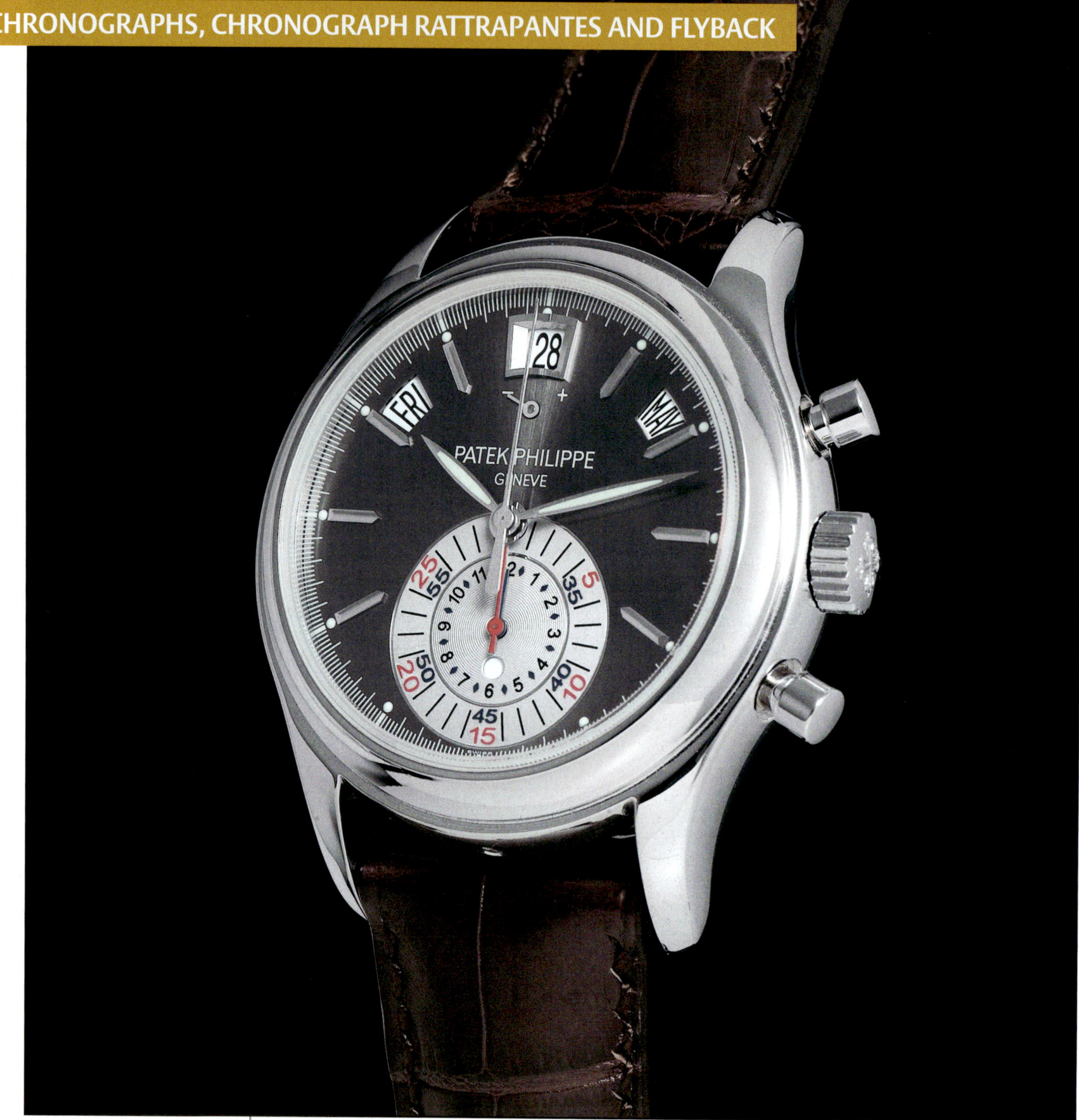


## PATEK PHILIPPE

## CHRONOGRAPH ANNUAL CALENDAR - REF. 5960P

This Chronograph Annual Calendar is crafted in platinum and houses the mechanical automatic-winding Patek Philippe CH28-520 IRM QA LU caliber, decorated with Côtes de Genève pattern and equipped with a 22K gold rotor. Hallmarked with the Geneva Seal, the watch offers hour, minute and annual calendar (date, day, month, moonphase) functions, and three chronograph counters. The case is set with a single brilliant at 6:00. The dial is solid gold with a sun pattern, luminescent leaf-style hands and blue and red enameled hour and minute counters. There is a day/night indicator at 6:00.

## PIAGET

## POLO CHRONOGRAPH

The Piaget Polo Chronograph model houses the new Calibre 880P, the first mechanical chronograph movement entirely designed, developed and produced by the Manufacture de Haute Horlogerie Piaget. In addition to the chronograph and flyback functions, this proprietary movement drives the hours, minutes and small seconds at 6:00, as well as the date display at 12:00 and a second time-zone indication. Within the select circle of manufacture-made chronograph movements, Calibre 880P stands out for its 24-hour dual time-zone display positioned at 9:00, the place usually reserved for the standard chronograph hour counter. Featuring two barrels and a large balance with screws, this mechanical self-winding movement provides a 52-hour power reserve in activated chronograph mode. Equipped with a column-wheel mechanism and vertical coupling-clutch, the 5.6mm-thick Calibre 880P is endowed with all the attributes of an haute horology chronograph movement.

## PIERRE DEROCHE

## SPLITROCK MDA

The SplitRock MDA concentric chronograph features several elements favored by Pierre DeRoche, including an individually numbered Dubois Dépraz movement, Côtes de Genève decorative motifs on all bridges, circular graining, blued-steel screws and an oscillating weight that is specially cut out. After its 2005 inauguration, the SplitRock collection gained a date display at 12:00, with a large window to ensure optimum readability. The completely damp-resistant rubber strap allows the wearer to sport the SplitRock MDA concentric chronograph in all conditions. It is available in polished and satin-brushed steel (MSRP $9,500) or 18K rose gold (MSRP $26,500).

## RICHARD MILLE

## RM 004-V2 SPLIT SECONDS CHRONOGRAPH

Representing a totally new design of split-seconds chronograph caliber with an unusual in-line escapement built on a three tiered titanium bridge, the manual-winding movement contains highly technical solutions to movement layout and functions. Built on a carbon nanofiber baseplate, it offers hours, minutes, seconds at 6:00, 30-minute counter, split seconds, power reserve and a torque indicator in dNmm. The column wheel and other parts of the movement are crafted from titanium; the exceptional movement can be seen through the sapphire crystal caseback. The unusual pushbutton layout was designed to optimize practical use of the chronograph and split-seconds function. The RM 004-V2 is offered in titanium, 18K red or white gold and platinum.

## RICHARD MILLE

## RM 011 FELIPE MASSA FLYBACK CHRONOGRAPH

This new Richard Mille creation offers a variety of functions such as hours, minutes, 60-minute countdown, seconds, flyback function and oversized date at 12:00 and a month indicator (numbered 1-12) at 4:00 with annual calendar only requiring correction for the month of February. Two unusually designed disks of cut out, metal digits are used for indicating the date. Using the skeletonized titanium, PVD-treated base caliber 005-S, the large case of 50x42x15.85mm has a central case-ring in titanium with the choice of both front and back bezel in 18K white or red gold.

## TAG HEUER

## 1911: THE TIME OF TRIP

This is the first dashboard chronograph, patented by Heuer in 1911, designed for aircraft and automobiles. With an 11cm diameter, it fits into all types of dashboards. The large hand at the center of the dial indicates the hour. The pair of small hands located at 12:00 gives the duration of the trip not exceeding 12 hours. The same pushpiece starts, stops and resets the watch to zero. A small window at 3:00 indicates whether the instrument is working correctly. Private collection of Tag Heuer museum.

## TAG HEUER

## 1916: THE MICROGRAPH

The Micrograph, the world's first stopwatch accurate to 1/100th of a second, was patented in 1916. This model revolutionized timing, notably in Olympic sprint competitions and technical research.
Private collection of Tag Heuer museum.

## TAG HEUER

## 1933: HERVUE PAIR

In 1933 Heuer began offering the Time of Day Clock called the "Hervue" and Twelve-Hour Timer called the "Autavia" mounted together on a double back-plate. In the late 1950s, Heuer began to refer to the Master Time clock and Monte Carlo stopwatch as the "Rally-Master" pair.
Private collection of Tag Heuer museum.

## TAG HEUER

## 1965: CARRERA DATE 45

After the launch of the Carrera series in 1964, Jack Heuer created the Carrera Dato 45 in 1965. It was one of the first chronograph models to feature a date disk (at 9:00). Previously, the date had been indicated by a hand.
Private collection of Tag Heuer museum.

## TAG HEUER

## 1975: THE CHRONOSPLIT

Chronosplit is the world's first quartz wrist-worn chronograph (with double digital display—LED and LCD; it is accurate to 1/100th of a second).
Private collection of Tag Heuer museum.

## TAG HEUER

## 2006: THE CARRERA CALIBRE 360 ROSE GOLD LIMITED EDITION (500 PIECES)

At BaselWorld 2005, TAG Heuer presented the Calibre 360 Concept Chronograph, the first mechanical wrist-worn chronograph to measure and display time to 1/100th of a second. At BaselWorld 2006, privileged collectors looking for the most accurate mechanical timepiece ever made had their first glimpse of the Carrera Calibre 360 Rose Gold, in a special limited edition of only 500 pieces.
Private collection of Tag Heuer museum.

## TAG HEUER

## 2006: MONACO CALIBRE 360 LS CONCEPT CHRONOGRAPH

At BaselWorld 2006, TAG Heuer introduced the next evolutionary phase of its 1/100th of a second chronograph technology: the Monaco Calibre 360 LS (Linear Second) Concept Chronograph, a daring new timepiece with an all-new architecture proudly exposing the unparalleled Calibre 360 LS precision technology within.
Private collection of Tag Heuer museum.

## TAG HEUER

## 2005: THE CALIBRE 360

The Calibre 360 beats 10 times faster than the most precise mechanical wristwatch movements on the market today. It is like being endowed with two hearts—2 sets of escapement mechanisms are allowed to cruise at a regular speed (28,800 oscillations/hour) under normal conditions then accelerate to immense speed (360,000 oscillations/hour) in chronograph mode. For a movement to be this incredibly precise, it requires more than 230 components, a lightened and miniaturized hairspring and escapement mechanism (26% smaller in diameter than an ordinary balance wheel), and, finally, no fewer than 3 exclusive worldwide patents: 1. 1/100th-of-a-second display on a mechanical wrist-worn chronograph; 2. Bi-directional crown and rewind system: The single crown controls the automatic watch and the mechanical chronograph, as well as the watch's hour and date setting. When turned counter-clockwise, it rewinds the automatic watch; when turned clockwise, it rewinds the manual chronograph; 3. Transmission of the date from the base of the movement to the upper dial.

## ULYSSE NARDIN

## MAXI MARINE CHRONOGRAPH - REF. 356-66-3/323

The Maxi Marine Chronograph, Ref. 356-66-3/323, is built in a stainless steel 41mm case and houses the UN-35 movement. The chronograph features an exclusive 45-minute register. It is available in 18K rose gold with various dial combinations and comes on either a bracelet or rubber strap.

## ULYSSE NARDIN

## MAXI MARINE CHRONOGRAPH - REF. 356-66-3/319

Crafted in 18K rose gold, the Maxi Marine Chronograph, Ref. 356-66-3/319, has a 41mm case that houses the UN-35 movement. The chronograph is equipped with the exclusive 45-minute register for diving. The watch is also available in steel and can be purchased on either a bracelet or rubber strap, and with various dial combinations.

## VACHERON CONSTANTIN

## MALTE CHRONOGRAPH PLATINUM LIMITED EDITION - REF. 47120

Created in a limited edition of just 75 pieces in 950 platinum, the Malte Chronograph houses the mechanical manual-winding Vacheron Constantin Caliber 1141 with column-wheel chronograph. It offers 48 hours of power reserve, is equipped with 21 jewels and beats at 18,000 vibrations per hour. The mainplate and parts are decorated with Côtes de Genève and circular-graining patterns, and it features some hand-finished component sides that are decorated with parallel lines and beveled. All steel parts and screw heads are beveled and hand polished. In addition to the hours, minutes and small seconds, the watch offers two chronograph counters. This Malte Chronograph is fitted with a curved antireflective sapphire crystal, sapphire caseback and is water resistant to 3atm. The dial is gray-grained platinum with hand-chased details and applied white-gold markers and hands. A member of the Excellence Platine Collection.

## VACHERON CONSTANTIN

## MALTE TONNEAU CHRONOGRAPH - REF. 49145

This 18K pink-gold Malte Tonneau Chronograph offers date indication via twin oversized apertures, in addition to the chronograph and time readouts. It is powered by the mechanical manual-winding Vacheron Constantin Caliber 1137 with 37 jewels and 40 hours of power reserve. The movement beats at 21,600 vibrations per hour. The chronograph offers a center seconds hand, 30-minute and 12-hour totalizers, and tracks to 1/6 of a second. On a hand-stitched alligator strap and pink-gold buckle.

## VACHERON CONSTANTIN
## OVERSEAS CHRONOGRAPH - REF. 49150/B01J

This 18K gold Overseas Chronograph houses the mechanical automatic-winding Vacheron Constantin Caliber 1137 with column-wheel chronograph. The movement beats at 21,600 vibrations per hour, is equipped with 37 jewels and offers 40 hours of power reserve. It is protected against magnetic fields by an inner iron case. Water resistant to 15atm, the watch offers hours, minutes, small seconds, large date and chronograph with three counters. In addition to a screw-down crown with case protector, the watch features a three-fold gasket and screw-down pushers. Its caseback is embossed with the Overseas sailboat and the hands and markers are luminescent.

## ZENITH

## CLASS EL PRIMERO

The Zenith Class El Primero houses the El Primero 4002 automatic chronograph movement that is a COSC-certified chronometer. With 337 components and 31 jewels, the movement beats at 36,000 vibrations per hour. It offers a power reserve of 50 hours and measures short time intervals to 1/10 of a second. In addition to hours and minutes in the center, it offers small seconds at 9:00, date at 4:30, central seconds hand and 30-minute counter at 3:00. It is crafted in a 40mm steel case with a curved sapphire crystal with antireflective treatment on both sides. The caseback is transparent sapphire, as well, and the watch is water resistant to 50 meters.

## ZENITH

## CLASS OPEN EL PRIMERO

Housing the El Primero 4021H automatic chronograph movement, this Class Open El Primero watch offers hours, minutes, small seconds, 50 hours of power reserve at 6:00, chronograph timing to 1/10 of a second, 30-minute counter at 3:00 and 3-branch seconds hand at 9:00. The 248-part movement beats at 36,000 vibrations per hour and has 39 jewels. Water resistant to 50 meters, this Class Open El Primero is crafted in steel and fitted with a sapphire crystal and transparent sapphire crystal caseback.

## ZENITH

## CHRONOMASTER GRANDE DATE

The ChronoMaster Grande Date houses the El Primero 4010 automatic chronograph movement with big date. The COSC-certified chronometer movement consists of 306 components, 31 jewels, beats at 36,000 vibrations per hour and holds 50 hours of power reserve. With bidirectional automatic winding, the watch is equipped with a central rotor on ball bearings and a 22K rose-gold oscillating weight. It measures to 1/10 of a second, has a 30-minute counter at 3:00, a 60-second counter at 9:00, and the big date at 6:00 features instantaneous double-jump function. A tachymetric scale rings the dial encased in 18K rose gold. The ChronoMaster Grande Date is water resistant to 50 meters.

## ZENITH

## CHRONOMASTER T

Housing the El Primero 410 automatic chronograph movement, this ChronoMaster T is a COSC-certified chronometer. The movement consists of 357 components and 31 jewels, beats at 36,000 vibrations per hour and retains up to 50 hours of power reserve. The chronograph measures to 1/10 of second. With bidirectional winding, the central rotor is on ball bearings and the oscillating weight features a grained pattern. In addition to the seconds and chronograph function with 30-minute and 12-hour counter, the watch offers day, date and month readout, moonphase at 6:00 and tachometric scale.

## ZENITH

## DEFY CLASSIC CHRONOGRAPH

Crafted in brushed stainless steel and water resistant to 300 meters, this Defy Classic Chronograph houses the El Primero 4000 SC automatic chronograph movement with 278 components. The balance bridge is crafted of shock-absorbing Zenithium Z⁺. The watch features an Incabloc anti-shock device and measures time to 1/10 of a second. The movement is built with a full Tungstentungsten oscillating weight with guilloché pattern, beats at 36,000 vibrations per hour and provides a power reserve of over 50 hours. Chronograph functions include 30-minute counter at 3:00, 12-hour counter at 6:00, and seconds at 9:00. This Defy Classic is equipped with a unidirectional rotating bezel and screw-in caseback.

## ZENITH

## DEFY CLASSIC OPEN

This Defy Classic Open houses the automatic El Primero 4021 SC chronograph movement with 248 parts, 39 jewels, 36,000 vph and 50-hour power reserve. Defy Classic Open offers 30-minute counter at 3:00, power-reserve indicator at 6:00, central chronograph seconds hand and a 3-branched seconds hand at 9:00. The steel case measures 46.5mm and is equipped with a unidirectional rotating bezel. Complete with transparent sapphire caseback, the watch is water resistant to 300 meters and accented with SuperLumiNova-tipped hands.

## ZENITH

## DEFY XTREME CHRONOGRAPH

The El Primero 4000 SX automatic chronograph movement powers this black titanium Defy Xtreme Chronograph. This caliber consists of 248 parts, 31 jewels and a heavy metal oscillating weight with carbon-fiber inserts. The central rotor is on ball bearings, the balance, chronograph and pallet bridges are of shock-absorbing Zenithium $Z^+$. The movement beats at 36,000 vibrations per hour and offers 50 hours of power reserve. The 46.5mm case is equipped with black titanium screw-in pushbuttons with carbon-fiber guilloché patterns, black titanium screw-in crown with special protective device, and a graduated unidirectional rotating black titanium bezel. The helium escape valve at 10:00 ensures water resistance to 1,000 meters. The dial is a multi-layered shock-resistant structure composed of silvered carbon fiber and displays a center chronograph hand, the 30-minute counter at 3:00, date at 4:00, 12-hour counter at 6:00, and small seconds at 9:00.

## ZENITH

## DEFY XTREME OPEN

Cased in 46.5mm of black titanium, this extraordinary Defy Xtreme Open is powered by the El Primero 4021 SX automatic chronograph movement with harmonic plate. The 13-lignes caliber consists of 248 parts and 39 jewels. The balance, chronograph and pallet bridges are built ofshock- shock-absorbing Zenithium Z+ and Defy Xtreme Open is equipped with a heavy metal oscillating weight with carbon fibercarbon-fiber inserts. It has a 30-minute counter at 3:00, small seconds at 9:00, power-reserve indicator from the hour axis, and central seconds hand. Water resistant to 1,000 meters, the high-tech case features a helium escape valve at 10:00 and black titanium screw-in pushbuttons with carbon-fiber guilloché patterns and protective crown device. The carbon-fiber dial is a multi-layered construction with Hesalite shock-resistant glass. The steel bracelet with Kevlar® inserts has links of carbon fiber and composite alloys for ultra resistance to extreme temperatures and pressure.

## ZENITH

## DEFY XTREME POWER RESERVE

Housing the Elite 685 SX extra-thin automatic movement, this Defy Xtreme Power Reserve watch offers power reserve readout at 1:00. Cased in a 43mm black titanium case, the movement is a 179-part caliber with 38 jewels beating at 28,800 vibrations per hour. The balance, automatic train and train wheel bridges are crafted in shock-absorbing Zenithium $Z^+$. The watch features an Incabloc anti-shock device. It offers 50 hours of power reserve. The sapphire crystal is 3.8mm thick with antireflective coating on both sides to enable water resistance to 1,000 meters. There is a helium escape valve at 10:00. The high-temperature-resistant bracelet is crafted in brushed steel with Kevlar inserts and carbon fiber and composite alloys for the central links.

## ZENITH

## GRANDE CHRONOMASTER OPEN EL PRIMERO

This Grande ChronoMaster Open houses the El Primero 4021 automatic chronograph movement with 249 parts. Beating at 36,000 vibrations per hour, the 39-jeweled movement measures short time intervals to 1/10 of a second. The watch has bidirectional automatic winding and holds more than 50 hours of power reserve, displayed from the center hour axis. A 3-branched small seconds hand is positioned at 9:00 and the 30-minute counter is at 3:00. Shown in stainless steel but also available in 18K rose gold, the 45mm case is water resistant to 30 meters and features a curved sapphire crystal with antireflective treatment on both sides and an exhibition sapphire crystal caseback. Available on alligator / Alzavel calfskin or rubber strap or stainless steel bracelet with triple-folding clasp; the dial is available in cobalt or white.

## ZENITH

## GRAND PORT-ROYAL EL PRIMERO CONCEPT

Cased in brushed stainless steel with a sapphire caseback and crystal, this Grand Port-Royal El Primero Concept is a stunning bidirectional winding automatic chronograph that houses the El Primero 4021 B movement with 248 parts and 39 jewels. The movement beats at 36,000 vibrations per hour and offers 50 hours of power reserve. It measures short time intervals to 1/10 of a second. Functions include 30-minute counter at 3:00, power-reserve display at 6:00, and 3-branched seconds hand at 9:00. The 36x51mm case is fitted with a silver or black guilloché dial and water resistant to 50 meters. Shown on a rubber strap, the watch is also available on a black alligator leather strap lined with silky Alzavel calfskin.

## ZENITH

## PORT-ROYAL OPEN EL PRIMERO CONCEPT

Crafted of black titanium with black titanium pushers and crown, this 34x48mm Port-Royal Open Concept houses the El Primero 4021 FN automatic chronograph movement consisting of 248 parts, 39 jewels, 50-hour power reserve and beating at 36,000 vibrations per hour; its heavy metal oscillating weight features a Damier guilloché pattern. The rectangular auxiliary plate is new and indications include 30-minute counter at 3:00, power-reserve display at 6:00, and 3-branched seconds hand at 9:00. On a carbon-fiber strap integrated with rubber and calfskin, the Port-Royal is water resistant to 50 meters.

# Multi-Complications

JEAN DUNAND
PIECE UNIQUE
SWISS MADE
MON
TUE
WED
THU
FRI
SAT
SUN
NOV
JAN
MAR
MAY
JUL
SEP
CHRISTOPHE CLARET
PIECE UNIQUE
MVT BY C. CLARET
REF.CLA 1879
750

## CEO of Audemars Piguet

# Georges-Henri Meylan

### Jules Audemars Tourbillon Minute Repeater Chronograph

The steady improvement of technologies and sciences often results from the competition between specialists of the disciplines concerned. In the case of watchmaking, the great names have historically proven able to provide wealthy connoisseurs with a positive image of their potential by producing watches with increasingly complex movements, year after year.

"With its three major complex mechanisms and in the current state of the horological art, it embodies the finest classic know-how cultivated by the manufacture."

This is certainly confirmed by Georges-Henri Meylan, CEO of Audemars Piguet, who proudly wears a watch that represents more than 130 years of history grounded in the mastery of the most sophisticated mechanisms. When asked why he wears a Jules Audemars Tourbillon Minute Repeater Chronograph, Meylan answers, "Quite simply because, with its three major complex mechanisms united within such a diminutive space, and in the current state of the horlogical art, it embodies the finest classic know-how cultivated by the manufacture." In a way, he says, it is "like carrying around with me the encapsulated essence of the brand." This timepiece certainly serves as a fantastic promotional tool, since the company has already sold around 30 of them within a year. "One must admit that this model, which is very different from the innovative Tradition d'Excellence N0 5, is reassuring for collectors who appreciate fine craftsmanship respectful of traditions."

*FACING PAGE*
*Georges-Henri Meylan, CEO, Audemars Piguet.*

*THIS PAGE*
*The Jules Audemars Minute Repeater Chronograph in its pink- gold version.*

### A response tailored to market requirements

Nonetheless, a watch creation may also serve to reflect trends within the international luxury market. Or, as Meylan puts it, "This Jules Audemars must enable the manufacture to stand out from the crowd and serve as a landmark by crystallizing the reactions of craftsmen who are tired of the increasingly burdensome presence of tourbillons

JULES
AUDEMARS
TOURBILLON
REPETITION
AUDEMARS PIGUET
SWISS
ADJUSTED TO HEAT
COLD AND 5 POSITIONS
AUDEMARS PIGUET
CHRONOGRA

FACING PAGE
*Center • The Jules Audemars Minute Repeater Chronograph features a rare combination of three complications.*

*Botton left • A sapphire crystal enables one to admire Caliber 2874.*

THIS PAGE
*The model also comes in a platinum version.*

presented in collections from generic brands." Moreover, above and beyond a straightforward reaction, "Audemars Piguet's very vocation lies in developing this type of construction." Representing an anthology of the expertise perpetuated by the talented specialists within the venerable firm based in Le Brassus that is still in the hands of the founding families, this watch instrument is crafted in pink gold with a black dial or in platinum with a silvered dial. "It unites three specialties that are particularly well mastered by the research and development unit centered in Le Locle," explains Meylan.

Representing a miniaturized engineering marvel, this mechanical hand-wound 13-lignes movement measuring 29.3mm in diameter and carrying reference number 2874, comprises 655 parts in all, including 38 jewels. At these dizzying peaks of excellence, each metal part receives individual treatment. Finishing details such as the mirror polishing (a surface treatment for hardened steel) and the beveling of the sharp ridges of the bridges, springs and levers—including those invisible to the naked eye—highlight the talent of the finishing craftsmen. By extension, as the CEO of the Audemars Piguet points out, "Each part requiring such meticulous care in being honed to perfection thus naturally draws its evocative force from the experience of the watchmakers responsible for individually assembling these components. Within such a context, the chronograph, a specialty that has become somewhat commonplace these days because of the sheer numbers being produced, regains its rightful pedigree within this resolutely Haute Horlogerie environment." In parallel, the tourbillon featuring a high-precision balance with dynamic adjustment screws is "transfigured" by the presence of a minute repeater mechanism that is seamlessly integrated within the caliber. Actually depressing the repeater slide thus allows this immaterial temporal concept to take on a genuinely physical dimension by being transformed into a supremely pure acoustic wave thanks to the presence of two gongs made from one piece and of hammers with improved operating capacities. "Quite obviously," admits Meylan, "announcing time in music naturally leads to even more complex mechanical approaches. On the basis of this watchmaking wonder with its crystal-clear tone, one can easily imagine an evolution towards a Grande Sonnerie or grand strike model, or a self-winding version of this exceptional movement." Nonetheless, while the future is undoubtedly important, Meylan appreciates creations in the present and patiently analyzes them in order "to ensure that each stage in the miniaturization of complexity consistently reinforces a legitimacy built up through 130 years of painstaking labor and guarantees it for the 130 years to come."

# Q & A with the President and CEO of Audemars Piguet North America

# François Benhamias

**Grand Complications:**
How has the American market evolved in recent years with respect to complicated watches?

**François Benhamias:**
The American market is experiencing tremendous growth as people become more and more educated about watchmaking and this growth has been most noticeable in the high-end category with significant interest in mechanical and complicated watches.

> "The American public is always interested in innovative products and technical features."

The complicated watches are highly sought after as clients come to understand they reflect the highest standard in craftsmanship and innovation in watchmaking. They exemplify what true luxury is and what the discerning American client has come to expect from Audemars Piguet. But this is still just the beginning.

**GC: What types of products specially interest American clients and connoisseurs?**

FB: The American public is always interested in innovative products and technical features. There is a huge interest in new materials and functions. Audemars Piguet is very committed to breakthrough innovation and this is a major asset in distinguishing us from other brands in the US market. The Millenary Cabinet No.5 with its new and revolutionary escapement and the Millenary MC12 with its interesting choice of materials and revolutionary design are two such examples of products that have resonated perfectly with the American client.

*FACING PAGE*
*François Benhamias, President e CEO of Audemars Piguet North America.*

*THIS PAGE*
*A miniature marvel of engineering and miniaturization, the mechanical hand-wound 13-lignes Caliber 2874 (29.30mm in diameter) driving the Jules Audemars Piguet Minute Repeater Tourbillon comprises 665 parts.*

AUDEMARS PIGUET

*FACING PAGE*
*This watch "is reassuring for collectors who appreciate fine traditional craftsmanship".*

*THIS PAGE*
*The Cabinet N° 5 is a Millenary Perpetual Calendar with Deadbeat Seconds and Power Reserve Indication.*

GC: What is your personal evaluation of the potential in the coming years of the US market, which is already the leading outlet for the Swiss watch industry?

FB: We are extremely optimistic because the level of knowledge and awareness is substantially increasing as the consumer becomes more discerning in terms of quality watchmaking. They start to notice what people wear on their wrists and it is becoming a statement as to who you are. If the American public gets to the level of watchmaking appreciation we already find in Italy, for example, we would never be able to supply enough watches to meet demand.

GC: How does Audemars Piguet plan to pursue its development in the American market? Which specific trump cards do you hold that will enable you to go further, higher... ?

FB: We are constantly striving to educate and connect with a younger clientele through partnerships like the one we have with the America's Cup Defender, Alinghi, and other unique projects in the sports and entertainment industry. This brings a new audience to the world of high-end watchmaking. The other major asset for Audemars Piguet is the uniqueness factor; the American client values the fact that we provide a very exclusive and iconic product—the true definition of luxury.

# Thierry Oulevay and Christophe Claret

## Join forces in the Jean Dunand Shabaka Minute Repeater and Perpetual Calendar

Shabaka, the latest Jean Dunand creation, represents a feat of micro-engineering complexity comprising a minute repeater chiming on cathedral gongs, an instantaneous jump perpetual calendar with a unique and original display of the dates, moonphases and leap-year cycle, as well as an ingenious state-of-wind indicator. Shabaka also has 4 patents and contains 721 components.

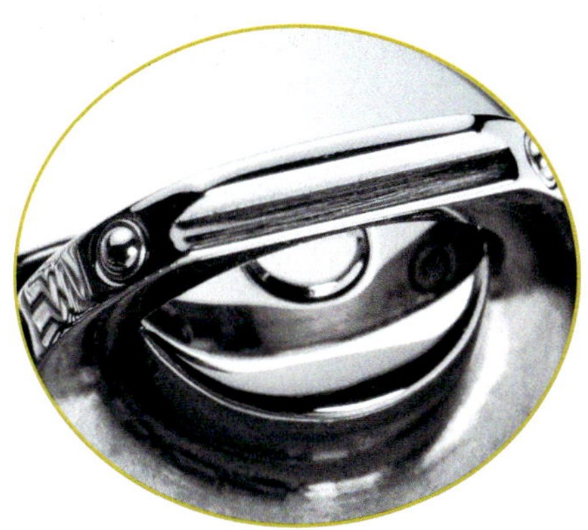

"Top-level technical innovation."

Designing and producing such an exceptional watch calls for a level of know-how that is the exclusive preserve of a select circle of watchmakers endowed with the necessary experience in the field of haute horlogerie. Jean Dunand is a youthful brand, but its founders are no newcomers to the scene. Behind the brand's creations are two well-known figures in the industry: Christophe Claret and Thierry Oulevay. Claret heads a complicated-movement manufacture based in Le Locle with a clientele comprising some of the most prestigious names in Swiss watchmaking; Oulevay, whose career included a period spent with Piaget, had revived the Bovet Fleurier brand before withdrawing from the firm.

For Oulevay and Claret, the very reason for heading the Jean Dunand brand is to pursue "top-level technical and aesthetic innovation and to offer unique models in the spirit of the 1920s and 1930s—an extremely ambitious objective in terms of quality." With this in mind, the company has chosen to work with the finest craftsmen (enamellers, engravers, etc.), all of whom play major roles in the specific nature and added value of such high-end watch products. Oulevay is firmly convinced of the complementary nature of small units operating alongside the great names in haute horlogerie. The market is proving him right. Youthful brands active in the high-end segment are successfully attracting the interest of connoisseurs with a taste for excep-

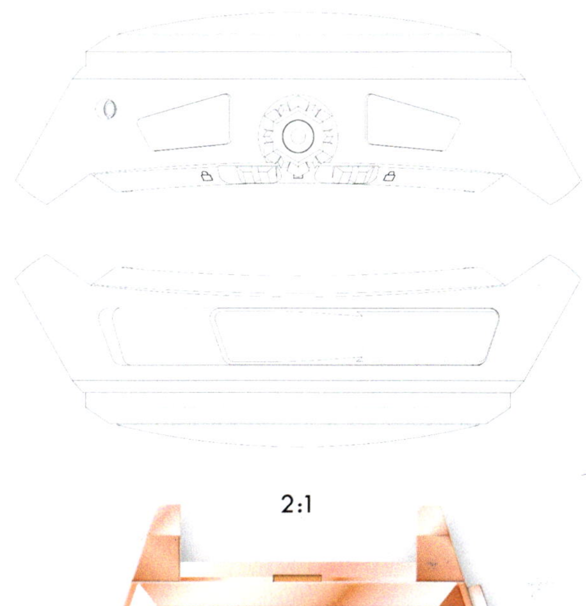

*FACING PAGE*
*Thierry Oulevay, CEO of Jean Dunand, with partner and movement maker extraordinaire, Christophe Claret.*

*This page*
*The back of the Shabaka in its pink-gold version.*

*Shabaka in its pink-gold version.*

*White-gold Shabaka model.*

*Back of the white-gold Shabaka.*

tional objects, not to mention the interest of collectors, who—in addition to their dozens of watches by established brands—are delighted to treat themselves to products that may be even more exotic, more exclusive, or more complex.

With the Shabaka Minute Repeater and Perpetual Calendar, Jean Dunand is once again set to intrigue genuine connoisseurs. The originality of this impressive timepiece is that its calendar indications are on four cylinders instead of the conventional disks. These are rotated by four different 90-degree transmission systems, each fitted with a security device to ensure precise calendar changes. Furthermore, unlike in most calendar indications, the date (two digits on separate cylinders), day and month jump instantaneously at midnight, when a permanently sprung mechanism is released.

The indication of the leap-year cycle is also novel: a white plate under the dial illuminates in turn the letter B (for bissextile) and the three ordinary years cut out on the dial between 7:00 and 8:00. Just opposite, in a perfectly symmetrical manner, the leap year is mirrored by the phases of the moon for which black disks travel over the surface of the moon like the shadow of the earth, eclipsing it to the left as it wanes and revealing it from the right as it waxes. The mechanical moon deviates from the real moon cycle by only one day every 120 years. For all the

*Two-tone white and pink-gold version of the Shabaka.*

complexity of the calendar indications, the state-of-wind indicator on the back of the watch is absurdly simple yet equally original. A single moving part—the mainspring itself in an open barrel—shows the power reserve of the watch against a graduated scale.

### Two gongs, two rotations

The minute repeater strikes on two cathedral gongs that are wrapped twice around the movement to give a deep, resonant chime. The minute-repeater slide on the left of the case means that the calendar setting controls are on the right. Two chronograph-style buttons are used to set the entire calendar (the gold button at 4:00) and the days only (at 2:00). A pushpiece set coaxially in the crown advances the months and years.

A complex series of levers reaches around the movement from the buttons to activate the respective calendar indications. The caliber is of a particular construction, with the perpetual-calendar mechanism integrated in the 13-ligne repeating movement. The 7mm diameter cylinders are embedded 2.5mm into the level of the minute-repeater to reduce further the thickness of the movement. An anthracite-black finish on the circular-grained baseplate and on the bridges decorated in concentric Côtes de Genève, contrasts with burnished steel, ruby and gold to highlight the beauty of the movement presented through the caseback.

### King of Egypt

The original calendar displays are integrated into the stunning geometrical design of the dial—a superb piece of engineering on four levels composed of a nickel-alloy frame delineating fields of blackened gold set with lapped pyramids of pink gold. Red ceramic hour-markers at 10:00, 12:00 and 2:00 serve to indicate the day, date and month respectively, against corresponding red ceramic triangles in the center of the dial.

The art deco-inspired design, with its Egyptian influences, is reflected in the exotic name of the watch—Shabaka, the 25th-dynasty pharaoh and king of Egypt.

As Claret explains, "Fundamentally, Jean Dunand is all about dreaming the impossible and then making it possible." This ambitious goal was clearly met by the first model presented by Jean Dunand, an orbital tourbillon representing a world-first achievement and a stunning technical challenge. The second was a grand complication watch with integrated mechanism, once again a pinnacle of the watchmaking art.

Given the specific nature and complexity of these timepieces, production is obviously strictly limited. No more than 20 Shabaka watches will be produced each year, thereby guaranteeing it a lasting aura of exclusivity.

*The Leroy 01 (1904) vied for many years with the Patek Philippe Graves watch for the title of the world's most complicated watch.*

# Multi-Complications

Ultra-complicated watches represent the very quintessence of horological expertise. With their movements comprising several hundred parts and their multi-display dials, multi-complications represent the culmination of an incredible amount of work combining tradition and innovation. Since the revival of the mechanical watch in the 1980s, the major manufactures—aided by cutting-edge computer technologies—have resumed the race for "the world's most complicated watch." Watchmakers also offer complex timepieces, comprising fewer functions but greater user-friendliness, functionality and readability. In short, a range of attractive assets that appeals to collectors and connoisseurs of rare models.

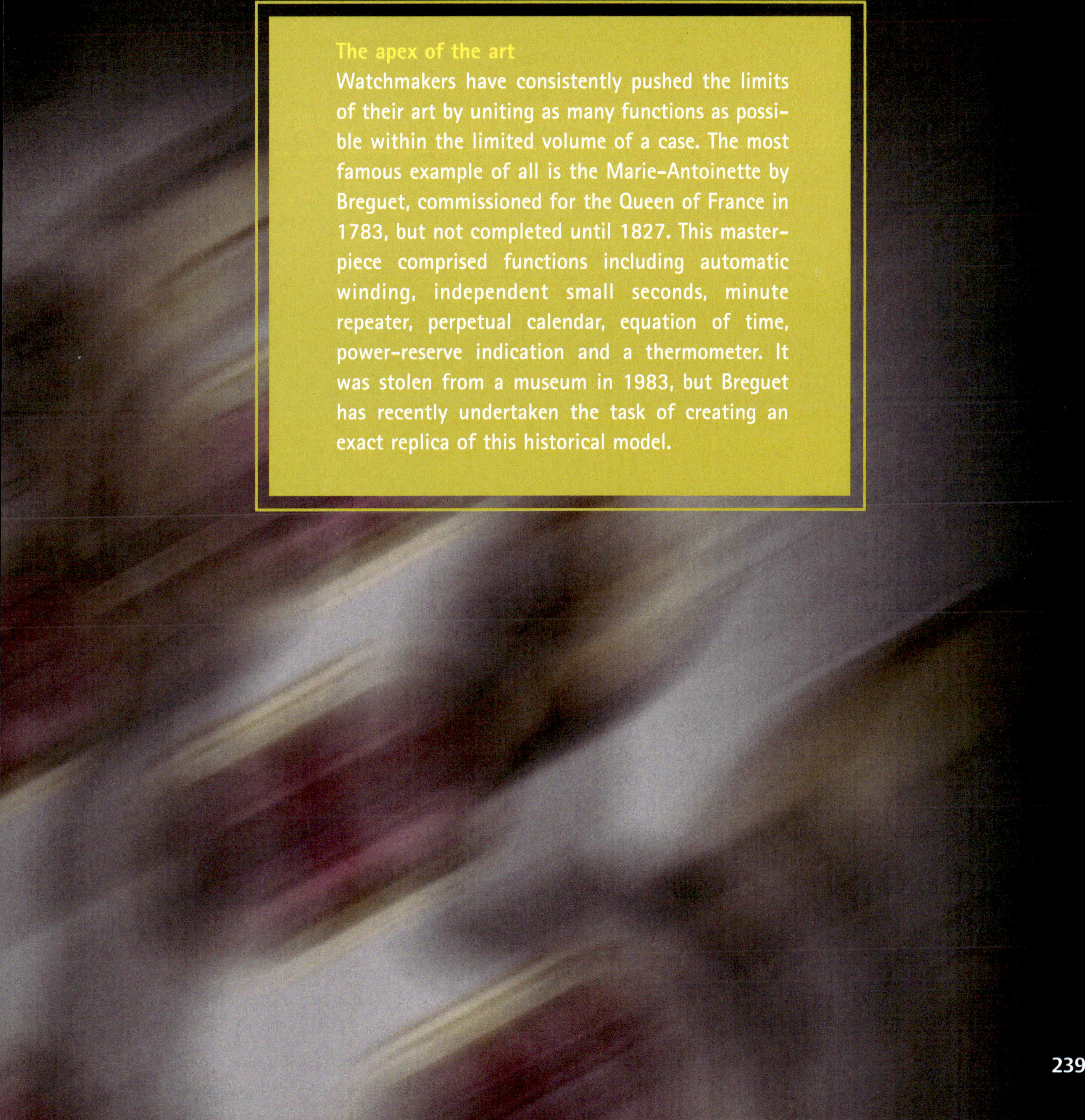

### The apex of the art

Watchmakers have consistently pushed the limits of their art by uniting as many functions as possible within the limited volume of a case. The most famous example of all is the Marie-Antoinette by Breguet, commissioned for the Queen of France in 1783, but not completed until 1827. This masterpiece comprised functions including automatic winding, independent small seconds, minute repeater, perpetual calendar, equation of time, power-reserve indication and a thermometer. It was stolen from a museum in 1983, but Breguet has recently undertaken the task of creating an exact replica of this historical model.

1

2

3

## A pocketful of records

If there is one particular name that stands out in the field of ultra-complicated watches, no doubt it is that of Patek Philippe. In 1925, the Geneva-based manufacture created a double-face pocket watch for Henry Graves, Jr. comprising 24 complications, including a grand strike mechanism chiming on four gongs as well as a map of the night sky as seen from New York City. For many years, the title of "the world's most complicated watch" was a source of debate among connoisseurs: was it the Graves watch, or the Leroy 01 (launched in 1904), which laid claim to between 20 and 25 complications. In 1989, Patek Philippe put an end to this ongoing dispute by presenting its

4

▸▸ 1 *Breguet, Marie-Antoinette*
*Breguet has set out to create an identical replica of the famous Marie Antoinette watch (created between 1783-1827) that was stolen in 1983.*

▸▸ 2 *and opening page (Leroy 01)*
*The Leroy 01 (1904) vied for many years with the Patek Philippe Graves watch for the title of the world's most complicated watch.*

5

▸▸ 3 *Patek Philippe, Graves*
*The twin-faced Graves pocket watch by Patek Philippe with its 24 complications was created in 1925 for New York banker Henry Graves Jr.*

▸▸ 4 *Patek Philippe, Calibre 89*
*With its 33 complications, the Patek Philippe Calibre 89 (1989) is still the most complicated portable watch in watchmaking history.*

▸▸ 5 *Patek Philippe, Star Caliber 2000*
*The twin-faced Star Caliber 2000 by Patek Philippe features the main astronomical functions on the twin dials.*

▸▸ 6 *Vincent Bérard, Quatre Saisons Carrosse*
*Vincent Bérard has reinterpreted the large "carriage" watches in his Quatre Saisons Carrosse.*

6

1

▸▸ 1 *Audemars Piguet, Grande Complication Savonnette*
*Grande Complication Hunter-watch by Audemars Piguet boasts a perpetual calendar, minute repeater and split-second chronograph.*

▸▸ 2 *Breguet, N° 5*
*A replica of an antique watch, the N° 5 by Breguet features a self-winding movement with a 60-hour power reserve, moonphases and "à toc" quarter repeater.*

▸▸ 3 *Vacheron Constantin, Tour de l'Ile*
*Equipped with 16 complications and composed of 834 parts, the Tour de l'Ile by Vacheron Constantin was unveiled in 2005 to mark the 250th anniversary of the Geneva-based manufacture.*

▸▸ 4 *Patek Philippe, Sky Moon Tourbillon*
*The Sky Moon Tourbillon wristwatch combines 12 rare and fascinating complications, including a minute repeater with "cathedral" chime and a mobile sky chart.*

2

Calibre 89 watch. Comprising 33 complications and 1,728 parts, this timepiece is still the most complicated wrist-worn watch in horological history. In addition to the classic complications, it also features some extremely rare functions such as the "secular" perpetual calendar (see Perpetual Calendars chapter) or the date of Easter. In 2000, Patek Philippe further distinguished itself by unveiling the double-face Star Caliber 2000 pocket watch with 21 complications. The goal in this case was not to bring together the largest number of functions, but rather "those representing time in the most poetic way," meaning first and foremost astronomical functions. Without aiming to enter the Guinness Book of Records, other watchmakers have recently presented ultra-complicated pocket watches endowed with an extremely sophisticated nature. Vincent Bérard has chosen to reinterpret

3

4

so-called "carriage clocks" by creating the Quatre Saisons Carrosse, equipped with five complications—including a perpetual calendar, a quarter-repeater mechanism and a system of automatons appearing on the dial—and graced with a grand feu enameled motif to match each of the seasons.

## Horological feats in wristwatch form

The emergence of the wristwatch in the early 20th century forced watchmakers to demonstrate a wealth of ingenuity in miniaturizing mechanisms and combining them within exceptional models. The race to create the world's most complicated watch is still a priority and a string of ultra-complicated watches has been launched with hopes of claiming the title. In 2001, Patek Philippe set the bar extremely high by presenting the Sky Moon Tourbillon, a double-face watch offering an original combination of the 12 complications regarded as the rarest and most fascinating, including a minute-repeater mechanism with cathedral chime, a tourbillon and a mobile celestial chart. In 2005, Vacheron Constantin celebrated its 250th anniversary by creating the Tour de l'Ile watch. The 834-part movement drives indications on two dials featuring 16 complications, including an hour-, quarter- and minute-repeater mechanism, tourbillon, display of the moon's phase

1

2

▸▸ 1 *Franck Muller, Aeternitas 8888*
*Franck Muller's Aeternitas 8888 houses 15 complications within an elegant tonneau-shaped case.*

▸▸ 2 *Jaeger-LeCoultre, Reverso Grande Complication à triptyque*
*With 18 complications displayed via three faces, the Jaeger-LeCoultre Reverso grande complication à tripytque stakes its own claim for the title of the world's most complicated watch.*

▸▸ 3 *Ulysse Nardin, Astrolabium Galileo Galilei*
*Ulysse Nardin's Astrolabium Galileo Galilei was the first model in the Trilogy of Time developed by Ludwig Oechslin.*

▸▸ 4 *Ulysse Nardin, Tellurium Johannes Kepler*
*The third member of Ulysse Nardin's Trilogy to Time tribute to astronomy is the spectacular Tellurium Johannes Kepler.*

▸▸ 5 *Ulysse Nardin, Planetarium Copernicus*
*Planetarium Copernicus is Ulysse Nardin's second piece in the Trilogy of Time.*

and age, perpetual calendar, equation of time, sunrise and sunset times and a sky map. In 2006, Jaeger-LeCoultre upped the ante by introducing its Reverso Grande Complication à triptyque, which makes unusual use of Reverso's famous swivel-case system created in 1931. This exceptional timepiece has three faces to interpret three dimension of time. The front shows civil time with a new ellipse isometer escapement and a titanium tourbillon; the back is dedicated to sidereal time with a sky map, zodiac calendar, equation of time and sunrise and sunset times; and the watch cradle or carrier displays perpetual time with a complete calendar—bringing the total complications for this watch to 18. Meanwhile, Franck Muller is hot on Jaeger-LeCoultre's heels with the Aeternitas 8888 uniting fifteen complications within an elegant tonneau-shaped case, including a so-called "secular" perpetual calendar, running equation of time, tourbillon and self-winding split-second chronograph function, along with a triple time-zone display.

### Reaching for the stars

Ulysse Nardin, a brand renowned for its marine chronometers,

has renewed ties with Renaissance astronomy and astrology in its Trilogy of Time. The first piece in the series, Astrolabium Galileo Galilei, has a dial displaying the functions of the astrolabe, a navigating instrument that led to the major discoveries in the age of the great explorers. In particular, it indicates the respective positions of the sun, the moon and the stars as observed from earth. The second member, Planetarium Copernicus, enables one to read at any time the month, current sign of the zodiac, and the astronomical positions of the planets in relation to the sun and earth by means of a dial composed of seven separate rings. Completing Ulysse Nardin's Trilogy of Time is Tellurium Johannes Kepler, which, with its cloisonné enamel dial depicting the earth viewed from above the North Pole, displays indications including sunrise and sunset times (with zones of shade and light) as well as solar and lunar eclipses. The whole world on the wrist!

6 Blancpain, 1735
*Within its 1735 watch, Blancpain united the six masterpieces of the watchmaking art: an ultra-thin movement, moonphases, perpetual calendar, split-second chronograph, tourbillon, and minute repeater.*

7 Jean Dunand, Grande Complication
*The Grande Complication by Jean Dunand (2005) houses a highly sophisticated mechanism developed by the brand's co-founder, Christophe Claret.*

## A. LANGE & SÖHNE

## DATOGRAPH-PERPETUAL - REF. 410.025

This model is driven by a manually-wound Lange movement: Calibre L952.1. It is a flyback chronograph with precisely jumping minute counter as well as perpetual calendar with outsize date, moon-phase display, day-of-week, month, leap-year display, small seconds hand with stop seconds, and day/night indicator. The plates and bridges made of untreated German silver, balance cock engraved by hand. The three-part platinum case features anti-reflection-coated sapphire-crystal glass and caseback, main pushpiece for simultaneously advancing all calendar displays as well as recessed pushpieces for separately advancing the calendar displays; two chronograph pushpieces. The hand-stitched crocodile strap is fitted with a solid-platinum buckle. The luminous two-part dial in rhodium-plated solid silver is equipped with a tachometer scale.

## A. LANGE & SÖHNE

## TOURBOGRAPH - REF. 702.025

The Tourbograph Pour le Mérite houses the manually wound Caliber L903.0 and is the world's first one-minute tourbillon in wristwatch format with a fusée-and-chain transmission, chronograph rattrapante functions, and power-reserve indicator. The mainspring and the fusée are interconnected with a delicate chain consisting of more than 600 parts. While the watch is being wound, the chain is wound up on the tapered fusée and the spring in the barrel is tensioned. The spring's power is delivered to the movement via the fusée, and thereby with constant torque. A planetary gear train composed of 38 parts keeps the movement running even while the mainspring is being wound. Without the chain, the movement consists of 465 parts and 43 jewels, two of which are diamond endstones. The three-part platinum case is equipped with sapphire crystal and caseback. It is created in a limited series of 51 pieces; a re-edition of 50 gold pieces is planned.

## AUDEMARS PIGUET

## JULES AUDEMARS GRANDE COMPLICATION CHRONOGRAPH RATTRAPANTE

## REF. 25984OR.OO.1138OR.01

Created in 18K pink gold, this Jules Audemars Grande Complication Chronograph Rattrapante features flyback function. Housing the self-winding Caliber 2885, which beats at 19,880 vibrations per hour, the watch is equipped with 52 jewels. The rattrapante enables split-second timing. In addition to chronograph functions and hours and minutes, the watch offers perpetual calendar functions with 52-week indication, astronomical moon and minute repeater. The watch features a sapphire caseback revealing the extraordinary movement. It is also available in yellow or white gold, or in platinum with sapphire caseback. Made to order.

## AUDEMARS PIGUET

## JULES AUDEMARS GRANDE COMPLICATION CHRONOGRAPH RATTRAPANTE

## REF. 25866PT.00.D002CR.01

The Jules Audemars Grande Complication houses the self-winding Caliber 2885 with 654 components. The watch offers perpetual calendar with 52-week indication, astronomical moon, minute repeater, and split-second chronograph. It is housed in a platinum case and features five correctors and a slide for the repeater on the side. The month and four-year cycle are positioned at 6:00 and the moonphase and date are shown at 12:00. This watch is also available in rose, yellow, white gold or platinum with sapphire caseback, on crocodile strap with double-blade AP deployment clasp in corresponding metal. Made to order.

## AUDEMARS PIGUET

## JULES AUDEMARS TOURBILLON CHRONOGRAPH - REF. 26010BC.00.D002CR.01

This Jules Audemars Tourbillon with Chronograph houses the exclusive Audemars Piguet tourbillon. The rhodium-plated movement features the Côtes de Genève pattern and circular graining. It offers chronograph function with 30-minute counter at 3:00 and a center sweep-seconds hand. The new hand-wound Caliber 2889 is equipped with 70-hour power reserve and the tourbillon has a progressive recoil click enabling fast winding of the watch. The titanium lever's low inertia eliminates 99% of the characteristic jumps made by the seconds hand when the chronograph is activated. This watch bears a sapphire crystal caseback and applied gold numerals riveted to the engine-turned dial. Its large-scale crocodile strap is secured with a gold AP folding clasp. This timepiece is also available in rose gold.

## AUDEMARS PIGUET

## MILLENARY MC12, LIMITED EDITION - REF. 26069PT.00.D028CR.01

Cased in 950 platinum, this tourbillon and chronograph houses the manual-winding Caliber 2884 that beats at 21,600 vibrations per hour and bears 30 jewels. This complication features a twin barrel tourbillon chronograph mounted in carbon base plate, secured by anodized aluminum bridges, cased in platinum with off-center displays, exposed by openwork dial. The Millenary MC12 displays hours and minutes, carbon 30-minute counter at 12:00, chronograph, and 10-day power-reserve indication. This edition is limited to 150 pieces and commemorates the partnership between Audemars Piguet and Maserati with the MC12.

## AUDEMARS PIGUET

### MILLENARY CABINET #5 - REF. 26066PT.00.D028CR.01

Millenary Cabinet #5 is the fifth inductee into the Tradition d'Excellence collection that will include 8 models. This 950 platinum timepiece is limited to just 20 pieces and each houses the manual-winding Caliber 2899 and features a world-premiere high-performance direct-impulse escapement equipped with 2 balance springs that requires no lubrication, twin barrels with 7 days of power reserve, linear perpetual calendar with adjustment through pushpiece and correctors recessed in the middle case, deadbeat seconds. Off-center displays, exposed by openwork dial with indications on multiple planes make the technical achievement a visual awe.

## AUDEMARS PIGUET

## POCKET WATCH GRANDE COMPLICATION CHRONOGRAPH RATTRAPANTE

## REF. 25712BA.00.0000XX.01

Cased in 18K yellow gold, the Classique Pocket Watch Grande Complication is powered by the manual-winding Audemars Piguet Caliber 2860 with minute repeater, perpetual calendar and split-seconds chronograph. The watch has a hunter-type design with a sapphire crystal and has been in production since 1875. Also available in platinum. Made to order.

## AUDEMARS PIGUET

## ROYAL OAK GRAND COMPLICATION - REF. 25865BC.OO.1105BC.01

The Royal Oak Grand Complication houses the self-winding Caliber 2885 and features a split-seconds chronograph, astronomical moon, a minute repeater and a perpetual calendar with day of the week indication. A subdial for the seconds is offered at 9:00. Cased in 18K gold, the movement beats at a steady 19,800 vibrations per hour. Twelve indicators are on the dial, including current year, leap-year indicator and moonphase readout. The movement is a labyrinth of 654 parts. Made to order.

## AUDEMARS PIGUET

## ROYAL OAK TOURBILLON CHRONOGRAPH - REF. 25977BA.001205BA.02

Crafted in yellow gold, this Royal Oak Tourbillon Chronograph houses the manual-wind Caliber 2889 with 70-hour power-reserve. The movement vibrates at 21,600 beats per hour and houses 25 jewels. The watch, with sapphire caseback, displays hours and minutes, subseconds dial and chronograph functions. This timepiece is also available in stainless steel.

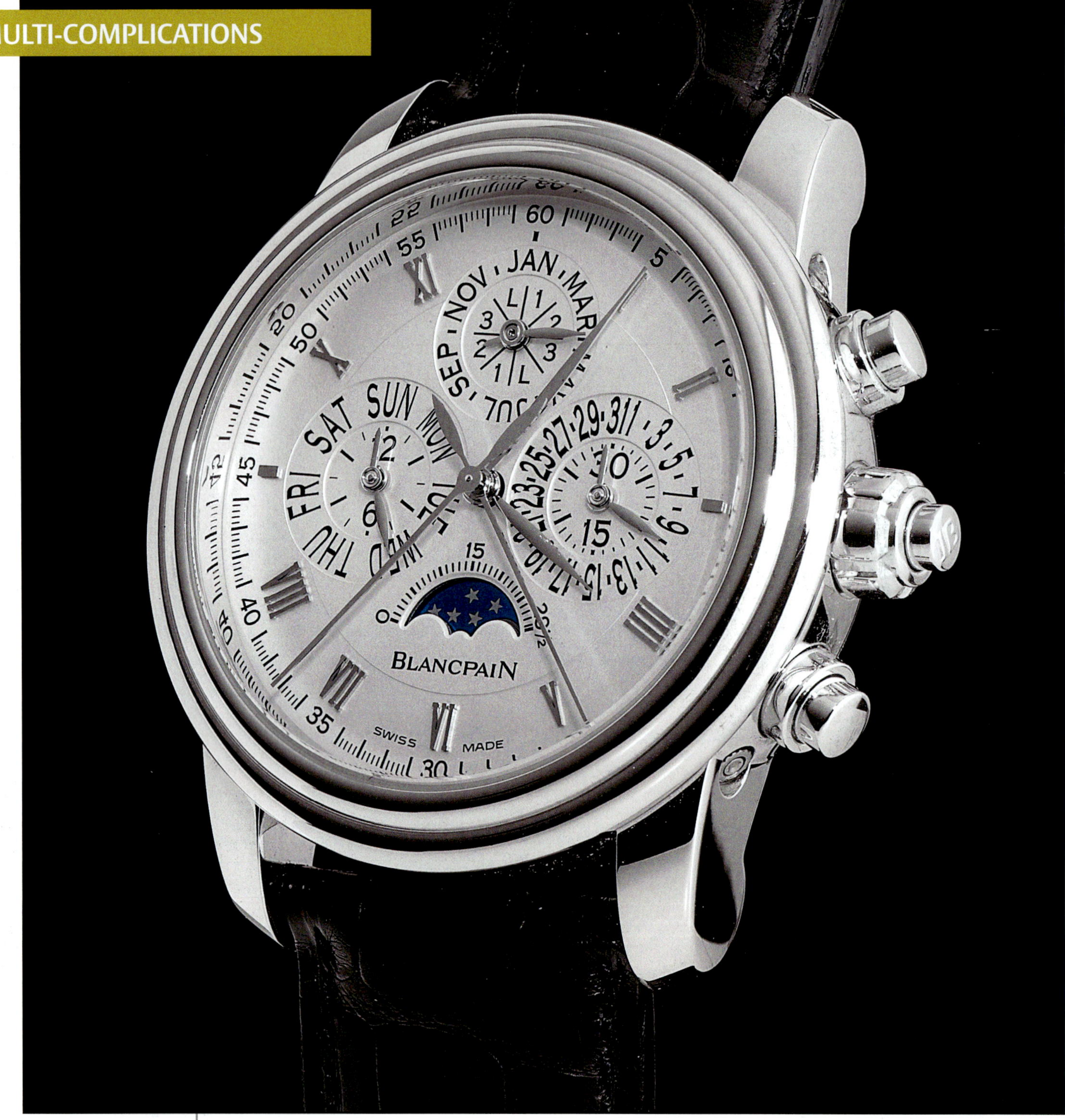


## BLANCPAIN

## LE BRASSUS SPLIT-SECONDS FLYBACK CHRONOGRAPH

This self-winding Blancpain caliber 56F9A (1185 base + 56QP module) offers perpetual calendar (date, day, month, four-year cycle and moonphase readout), and flyback split-second chronograph. This limited-edition model has a platinum case and closed back in an extra-large size (Ø 42mm). On the opaline dial, calendar and chronograph information appears in an analog big-size display assuring the best possible readability. The chronograph is provided with an additional split-second hand.

## BLANCPAIN
## QUATTRO

Created in the Le Brassus case, the Quattro houses a tourbillon, flyback split chronograph, and a perpetual calendar. The 42mm model is offered in platinum or in 18-karat rose gold. Only 50 pieces will be made in platinum.

## BLANCPAIN

### VILLERET PERPETUAL CALENDAR WITH CORRECTORS HIDDEN UNDER THE LUGS

Blancpain clothes its Perpetual Calendar self-winding watch within an original, practical and attractive technical innovation featuring correctors hidden under the lugs. The result of this in-house development is protected by a Blancpain patent. Made up of 297 parts and endowed with a 100-hour power reserve, self-winding Calibre 5653A drives the various displays to indicate the hours and minutes, day of the week, date, month and moonphases, as well as the leap year. Water resistant to 30 meters, the new Villeret Perpetual Calendar with correctors hidden under the lugs is available in a 38mm case crafted in 18K red gold or in platinum in a limited edition of 201 pieces.

## BREGUET

## CLASSIQUE GRANDE COMPLICATION – REF. 1907BA/12

This Classique Grande Complication pocket watch is crafted in 18K yellow gold and features a grand strike and tourbillon. The hand-engraved and hand-wound movement has a 2-way rotating crown and features an off-centered chapter ring and center minute hand on a silvered gold dial, hand-engraved on a rose-engine. The caseback is protected by a sapphire crystal.

## BREGUET

## CLASSIQUE GRANDE COMPLICATION - REF. 3755PR/1E/9V6

This Classique Grande Complication open-worked wristwatch is crafted in a platinum water-resistant case. Its hand-wound movement in pink gold is engraved by hand and features include a tourbillon, with small seconds on the tourbillon shaft, and a perpetual calendar showing the day, date, month and leap years, and a compensating balance-spring with Breguet overcoil.

## BREGUET

## MARINE - REF. 5837BR/92/5ZU

This Marine tourbillon with chronograph is crafted in 18K rose gold and houses a hand-wound movement, featuring a tourbillon with small seconds on the tourbillion carriage. Its black rhodium-plated 18K gold dial, engine-turned in a wave pattern, displays a chapter ring with pink gilt applied Roman numerals and luminous dots and a 12-hour totalizer at 6:00. Its Breguet hands are luminous, open-tipped and 18K rose gold. Its case is enhanced with a sapphire caseback and complemented with a rubber strap with folding clasp.

## BREGUET

## CLASSIQUE GRANDE COMPLICATION - REF. 5327BR/1E/9V6

This Classique Grande Complication wristwatch is crafted in 18K yellow, white or rose gold, and houses a thin, engraved automatic movement with perpetual calendar, precise moonphases, 46-hour power reserve and 3Hz balance frequency. Its engine-turned dial is in silvered 18K gold with a chapter ring displaying Roman numerals, central months indication, and showing dates, days and leap years on three subdials, power-reserve indicator at 10:30 and moonphases at 1:30. It also features a secret signature and power-reserve indicator. The engine-turned caseback is sapphire crystal and the case is water resistant to 30 meters.

## BREGUET

## CLASSIQUE GRANDE COMPLICATION - REF. 5347PT/11/9ZU

This Classique Grande Complication wristwatch is crafted in platinum and houses a hand-wound movement, featuring a center plate engine-turned by hand including two apertures for the tourbillons, which rotates in step with the passing hours, and Breguet overcoils. Its ring-shaped dial forms a flange in silvered 18K gold and displays a chapter ring with Roman numerals. The hour hand is an extension of the bridge supporting the two tourbillon carriages. The water-resistant case has a sapphire caseback, bearing a hand-engraved drawing representing the solar system.

## BVLGARI

## ASSIOMA MULTICOMPLICATION - REF. AA48PLTB

The Assioma Multicomplication features an in-house manufactured, mechanical automatic-winding Caliber BVL 416 with tourbillon device. The movement is finished and decorated by hand with stippling. It also offers a white-gold skeletonized rotor with applied yellow-gold engraved ring and hanging tourbillon bridge. The 48mm case is crafted in curved polished platinum and is fitted with a curved antireflective scratch-resistant sapphire crystal and a sapphire crystal snap-on exhibition caseback. The in-house manufactured dial with minute circle, GMT display and date counter is decorated with guilloché, clou de Paris and striated motifs. The pre-curved rolled-edge brown alligator strap closes with a platinum triple-fold-over buckle. Limited edition of 25 pieces.

## GERALD GENTA

## ARENA TOURBILLON PERPETUAL CALENDAR MOON PHASE

## REF. ATM.X.75.860.CN.BD

The Arena Tourbillon Perpetual Calendar Moon Phase features a Gérald Genta exclusive tourbillon automatic manufacture movement, offering a 64-hour power reserve. It is coated in antique gold and decorated with a concentric circular-graining. It is housed in a Ø 41mm platinum case with fluted caseband and palladium bezel. The black, red and satin-finish metallic skeleton dial offers original indications for moonphase, leap years and months, days of the month, small seconds and days of the week.

## GERALD GENTA

## OCTO TOURBILLON PERPETUAL CALENDAR WWT - REF. OTW.Y.50.930.CN.BD

The Octo Tourbillon Perpetual Calendar Moon Phase World Wide Timer offers an entirely hand-decorated Gérald Genta in-house tourbillon movement with perpetual calendar, phases of the moon, and global time on a rotating bezel (WWT). It is housed in a Ø 42.5mm octagonal platinum case with tantalum bezel, enhanced by a white-gold guilloché and cloisonné ivory and black ceramic dial and beaded crown with Falcon-eye cabochon. This model is water resistant to 10atm and is also available in a red-gold case with a red-gold guilloché and cloisonné black and red ceramic dial.

## GIRARD-PERREGAUX

## CLASSIQUE ELEGANCE WW.TC QUANTIÈME PERPETUEL - REF. 90280

This Classique Elegance ww.tc Quantième Perpetual is powered by the mechanical automatic-winding GP 033Q0 caliber with 46-hour power reserve and 26 jewels. Beating at 28,800 vibrations per hour, it is beveled and decorated with a Côtes de Genève pattern. The anthracite dial offers hours, minutes, perpetual calendar (date, day, month, four-year cycle), world time, and 24-hour indication via a white and black 24-hour ring, luminescent applied pink-gold hands and markers, and blued-steel secondary hands. The world time is read from an intermediate ring (turning together with the main hour) with 24 hours divided by day and night (white-black), and a turning flange with reference towns for the 24 time zones. Water resistant to 3atm this watch's three-piece, 41mm 18K pink-gold case has an antireflective curved sapphire crystal, screw-down crowns and screw-down caseback.

## HARRY WINSTON

## WESTMINSTER TOURBILLON

Issued as part of the Ocean Collection, this exceptional Westminster Tourbillon combines the tourbillon and the minute repeater, which chimes the same notes as the clock tower on the Palace of Westminster. The watch houses four cathedral gongs and is equipped with a flying tourbillon. Beating at 21,600 vibrations per hour, the exquisite 34-jeweled movement offers 70 hours of power reserve and is housed in a 45mm 18K rose-gold case.

## HUBLOT

## BIGGER BANG - REF. 308.TM.130.RX

Launched in 2006, the Bigger Bang is dedicated to Hublot's Art of Fusion concept, merging the traditional art of watchmaking and the visionary art of 21st-century watchmaking. Its exceptional HUB 1400 CT movement is the first column-wheel chronograph with direct coupling on a flying tourbillon cage. Shown here with a 44.5mm platinum PT950 case and matte black ceramic bezel on a natural structured adjustable black rubber strap, each Bigger Bang is water resistant to 50 meters. Each openwork dial reveals the chronograph driven from the 1-minute rotating tourbillon carriage, which is raised 2.8mm for maximum visibility as the movement's masterpiece. This model is limited to 18 pieces.

## IWC

## GRANDE COMPLICATION - REF. 377019

This self-winding IWC caliber 79091 (Valjoux 7750 modified base + IWC calendar and repeater modules) offers perpetual calendar (date, day, month, four-digit year, moonphase), minute repeater, and chronograph with three counters. The most impressive complications—perpetual calendar, repeater and chronograph—are combined in this historical watch, a highly valuable timepiece that allows automatic calendar updating until the year 2499; the digits representing 2200 to 2500 are stored in a small glass tube, supplied with the watch. The four-digit year display and the new repeater system with a reduced number of mobile components (some of which have special new designs) are witnesses of a superior class. The Ø 42mm case is waterproof and antimagnetic.

## IWC

## PORTUGUESE PERPETUAL CALENDAR - REF IW502119

This 18K rose-gold Portuguese Perpetual Calendar watch is powered by the company's 5000 caliber and offers ingenious displays. It houses the Pellaton winding mechanism with an entirely new moonphase display, a world-first patent. The Portuguese Perpetual Calendar offers seven days of power reserve and all perpetual calendar readouts. The moonphase is represented in the sky's northern and southern hemispheres, as well as via a disk display with two opposing circular windows rotating above a yellow surface with two black circles. Both moons are constantly in motion.

## IWC

## GRANDE COMPLICATION - REF. 377015

This Grande Complication offers chronograph, perpetual calendar, perpetual moonphase indicator, four-digit year indicator, minute repeater and small seconds with stop device. It houses the caliber 79091 movement with 75 jewels and 44 hours of power reserve. The watch beats at 28,800 vibrations per hour.
ALTERNATE VERSION: 9270 IN PLATINUM WITH PLATINUM BRACELET

## JAEGER-LECOULTRE

## GYROTOURBILLON I

Jaeger-LeCoultre's patented GyroTourbillon I houses the Caliber 177 complete with a built-in running equation-of-time mechanism, as well as a host of other functions and indications. The exquisite escapement is a first in the history of the tourbillon due to its spherical design, whose technically advanced multi-dimensional rotation is fascinating. The complex caliber is comprised of 512 parts. In addition to depicting the hours and minutes, its dial also indicates the remaining power reserve, date, month and true solar time as part of the equation-of-time function. The perpetual calendar will not require manual intervention until February 28, 2100. The spherical tourbillon features an aluminum case and an inner titanium and aluminum carriage that is oriented at a 90-degree angle to the outer case and rotates 2.5 times faster than its companion. The total weight of the nearly 100 components that comprise this spherical wonder is just 0.33 grams.

## JAEGER-LECOULTRE
## MASTER TOURBILLON

Clad in a traditional round pink-gold case, the Master Tourbillon is powered by the automatic Jaeger-LeCoultre Caliber 978 beating at 28,800 vibrations per hour and endowed with a 48-hour power reserve. It features a relief-engraved 22K gold monobloc oscillating weight. Its sunray-brushed dial with gold-plated numerals and hour-markers features central hour and minute hands, seconds, a dual time zone and an innovative date display around the dial rim. The latter jumps from 15 at 4:30 to 16 at 7:30 in order to avoid disturbing the serene contemplation of the tourbillon at 6:00. Water resistant to 50 meters, this Master Tourbillon comes on an alligator leather strap with triple-blade folding clasp.

## JAEGER-LECOULTRE

## REVERSO GRANDE COMPLICATION À TRIPTYQUE

This Reverso model displays three dimensions of time on three faces: civil, sidereal and perpetual time, made possible by a specially adapted, manually wound mechanical Jaeger-LeCoultre movement. The watch's front dial has a tourbillon endowed with an ellipse isometer escapement, which does away with the traditional Swiss pallet. Its back dial features a zodiacal calendar with an astronomical chart, while the third dial is remarkable for its unique location: a perpetual calendar in the Reverso carriage. A highly original, patented mechanism makes it possible to transmit energy from the movement to the carriage. In total, the timepiece boasts 18 complications, and has six patents pending.

## JEAN DUNAND

## SHABAKA

The Shabaka minute repeater and perpetual calendar with moonphases is Jean Dunand's impressive new Pièce Unique, offering a dramatic display of calendar information. Cylinders on 90-degree transmissions show dates, days and months that advance instantaneously at midnight. The leap-year cycle is indicated by a white plate illuminating cut-out figures on the dial, while the state of wind is revealed ingeniously by exposing the mainspring against a scale. The repeater chimes on cathedral gongs that circle the movement twice. The displays are integrated in the Art Deco design of the dial, engineered on four levels and set with pyramids of gold. Produced by Thierry Oulevay and manufactured by Christophe Claret, the Shabaka takes its name from the Pharaoh of Egypt's 25th Dynasty.

## PARMIGIANI FLEURIER

## TORIC CORRECTOR MINUTE REPEATER AND PERPETUAL CALENDAR

A world premiere timepiece, this Toric houses the mechanical manually wound Parmigiani Fleurier caliber with retrograde perpetual-calendar module entirely crafted in-house. The 33-jeweled movement vibrates at 21,600 beats per hour and offers 45 hours of power reserve. The movement consists of hand-beveled bridges and the Côtes de Genève decorative motif. It offers hours, minutes, day, retrograde date, month, leap year and a high-precision moonphase indicator, as well as the minute-repeater mechanism that strikes on two gongs. The watch is offered in platinum or 18-karat pink gold with a pushpiece that instantly corrects all perpetual calendar and moonphase functions. The dial is silvered 18-karat gold with barely corn guilloche' motif.

## PATEK PHILIPPE

## CHRONOGRAPH PERPETUAL CALENDAR - REF. 3970E

This model's manual-winding movement with Geneva Seal (Patek Philippe caliber CH 27-70 Q, CH 27-70 base + calendar module) offers perpetual calendar (date, day, month, four-year cycle, moonphase) and chronograph with two counters.

## PATEK PHILIPPE

## PERPETUAL CALENDAR MINUTE REPEATER - REF. 5074

This self-winding, 467-part movement with Geneva Seal (Patek Philippe caliber R 27 Q, R 27 PS base + calendar module) offers perpetual calendar (date, day, month, four-year cycle, moonphase) and minute repeater. The Ø 40mm case houses the repeater-perpetual movement realized by Patek Philippe long ago by combining the automatic base including a minute repeater (R 27 PS caliber without seconds) and the most classic calendar module with 24-hour indications. This model, presented in 2001 with only with the repeater, is meant for connoisseurs. The case size assures a particularly clean amplification of the Cathedral sonnerie. The gong is realized with a special alloy and undergoes several rotations. The full and long-lasting sound thus obtained recalls the bells of ancient cathedrals. The special alloy used for the acoustic gongs, which assures an extraordinary resonance, was developed by Patek Philippe in cooperation with the metallurgy experts of the Federal Swiss Institute of Technology of Lausanne.

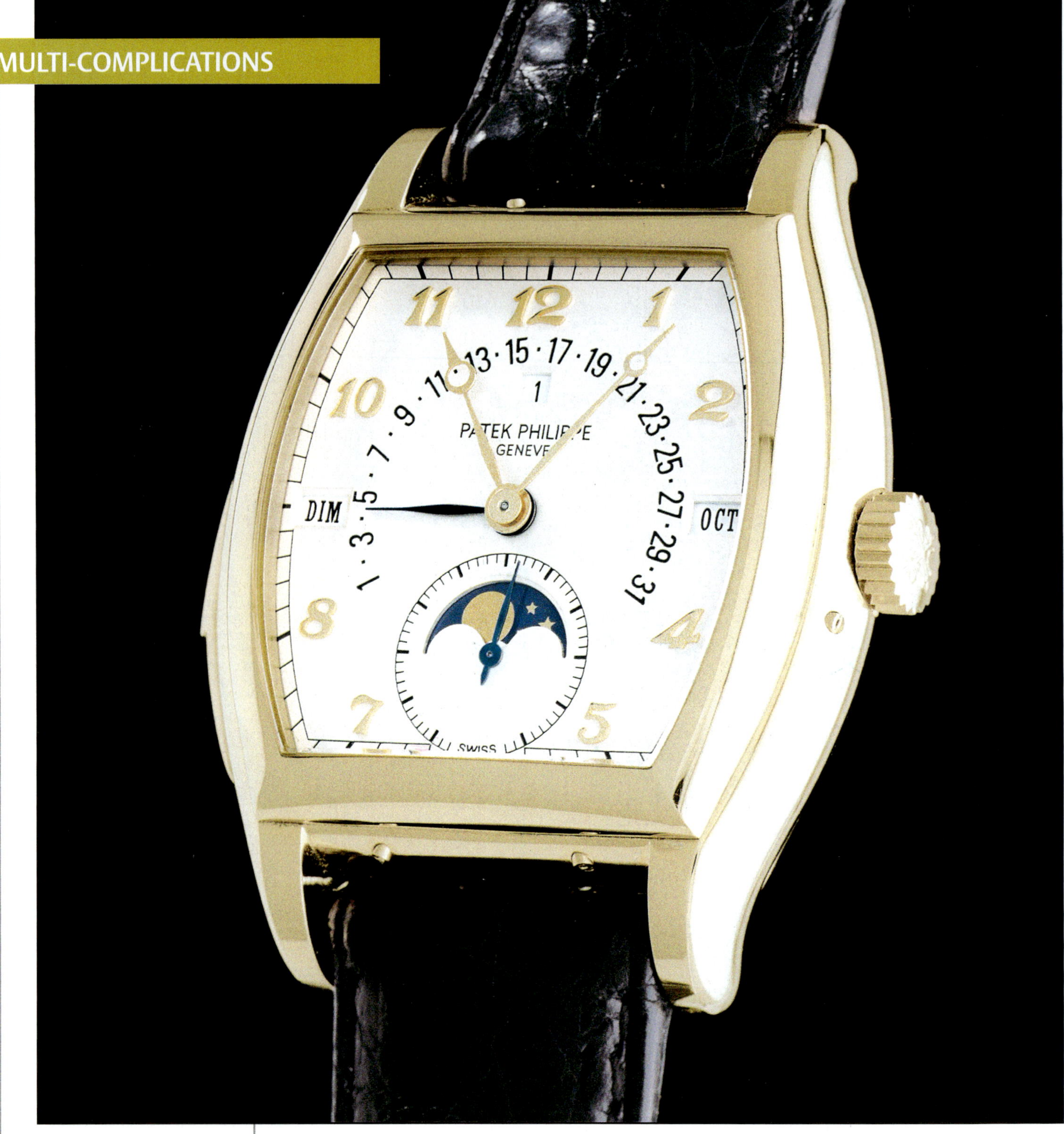


## PATEK PHILIPPE

## PERPETUAL CALENDAR REPEATER TONNEAU - REF. 5013

This self-winding movement with Geneva Seal (Patek Philippe caliber R 27 PS QR, R 27 PS base + 126 calendar module) offers perpetual calendar (retrograde date, day, month, four-year cycle, moonphase) and minute repeater. The elegant tonneau-shaped case with sapphire crystal caseback has a repeater slide and four small, integrated pushers on its middle for fast calendar correction. The dial shows the analog indications of hour, minute, small second and date (the latter with a retrograde hand), while the day of the week, the month and the four-year cycle are digitally displayed in small windows.

## PATEK PHILIPPE

## SKY MOON

This self-winding movement with Geneva Seal (caliber 240 CL LU, 240 base + 165 module) offers a sky chart and moonphase. Without further complications, but provided with the thin 240 automatic movement with its integrated micro-rotor, Patek Philippe's Sky Moon model reproduces the portion of the sky visible from a certain place (in this piece, the city of Geneva) enclosed in an ellipsis (representing the horizon) on the main dial. It indicates the positions and sidereal hours of Sirius and the Moon, as well as the moonphase in an additional window. These indications are displayed by sapphire disks controlled by a complexity of gears totaling 301 components. The Sky Moon Tourbillon's astronomic mechanism (686 components) is protected by a patent.

## PATEK PHILIPPE

## SKY MOON TOURBILLON - REF. 5002J

This manual-winding movement with Geneva Seal (Patek Philippe caliber RTO 27 QR SID LU CL) offers tourbillon, minute repeater, perpetual calendar, (retrograde date, day, month, four-year cycle, moonphase) and astronomic functions. On the sky map, the moon (with phases) and the Sirius star's sidereal positions are indicated on a 24-hour scale. The sidereal day is determined by two subsequent passages of a star on a meridian (here, the one of Geneva). The sidereal time indication is useful to determine the longitude from a certain place. The ellipsis moves and displays, at any moment, the part of the sky visible from that place (in this case, Geneva). The variations of astronomic indications by a few seconds each century are the slightest ones ever obtained by a mechanical watch.

## PATEK PHILIPPE

## SPLIT-SECOND CHRONOGRAPH PERPETUAL CALENDAR - REF. 5004

This manual-winding movement (Patek Philippe caliber CHR 27-70 Q, CHR 27-70 base + calendar module) offers perpetual calendar (date, day, month, four-year cycle, moonphase) and split-second chronograph. In 1995, after years of research and development, Patek Philippe's technicians and watchmakers brought one of its most famous complicated wristwatches—realized during the 1950s as Reference 2571—to new life: a split-second chronograph with two counters and a perpetual calendar, moonphase indications and a 24-hour display. Its modern descendant, characterized by the digital day and month displays and by the split-second pusher placed on the crown. The gold or platinum case is provided with two backs: one closed and one transparent.

## RICHARD MILLE

## RM 008-V2 TOURBILLON SPLIT SECONDS CHRONOGRAPH

Representing a totally new design of spilt-seconds chronograph movement with tourbillon escapement, this watch was the result of many years of development. Built on a carbon nanofiber baseplate, it offers hours, minutes, seconds at 6:00, 30-minute counter, split seconds, power-reserve indicator and a torque indicator in dNmm. The column wheel and other parts of the movement are crafted from titanium; the exceptional movement can be seen through the sapphire glass caseback. The unusual pushbutton layout was designed to optimize practical use of the chronograph and split seconds function. The RM 008-V2 is offered in titanium, 18K red or white gold and platinum.

## VACHERON CONSTANTIN

## MALTE CHRONOGRAPH PERPETUAL CALENDAR - REF. 47112

The manual-winding movement (Vacheron Constantin 1141 QP, Lemania base + calendar module) offers perpetual calendar (date, day, month, four-year cycle, moonphase) and chronograph. The modifications Vacheron Constantin made to the basic movement, produced by Nouvelle Lemania and visible through the wicket-hinged caseback, refer particularly to the shapes of some bridges, the personalized decorations and the addition of a perpetual-calendar module. The moonphase is indicated on a gold disk engraved by hand. The four-year cycle (consisting of the three digits 3-6-5, indicating the normal year, and 3-6-6 for the leap year) is displayed through the unusual window system.

## VACHERON CONSTANTIN
## MALTE MOONPHASE - REF. 83500

Available in either 18K white or yellow gold, this Malte Moonphase houses the Caliber 1410 manually wound movement with 40-hour power reserve, the Hallmark of Geneva, and beating at 28,800 vibrations per hour. The dial is engine-turned and 82 brilliant-cut diamonds weighing approximately 0.71 carat adorn this watch. Water resistant to 3atm.

## VACHERON CONSTANTIN

## PATRIMONY TOLEDO 1952 - REF. 47300

This Patrimony Toledo 1952 is crafted in 18K pink gold and houses the mechanical manual-winding Vacheron Constantin Caliber 1125. Beating at 28,800 vibrations per hour, the complicated movement is equipped with 36 rubies and offers 40 hours of power reserve. In addition to small seconds, the watch offers a full calendar (date, day, month, moonphase). The three-piece square case is fitted with a curved sapphire crystal and is water resistant to 3atm. On a hand-stitched alligator strap with pink-gold clasp.

## VACHERON CONSTANTIN

## PATRIMONY TRIBUTE TO GREAT EXPLORERS - REF. 47070

Crafted in 18K yellow gold, these Patrimony Tribute to Great Explorers watches house the self-winding Caliber 1126, which boasts 36 jewels, beats at 28,800 vibrations per hour and offers 38 hours of power reserve. Their enameled two-piece dials feature mobile hour tracks that span across 120-degree arcs. Patrimony is water resistant to 30 meters.

## VACHERON CONSTANTIN

### ST. GERVAIS

This complicated watch pushes the existing limits of the power reserve. A world-first with four barrels coupled with a regulator tourbillon and a perpetual calendar, the watch has 250 hours of power reserve. The 44mm platinum-cased timepiece houses the new caliber 2250 with 410 parts. In addition to the gold hour and minute hands, two blued steel hands run over the silvered engine-turned gold dial in arcs at 3:00 and 9:00 to indicate the power reserves of the barrels. Only 55 pieces of this watch will be created.

## VACHERON CONSTANTIN

## TOUR DE L'ILE

Vacheron Constantin's master watchmakers invested more than 10,000 hours on research and development—or almost two years—into this multi-complication. Only seven Tour de L'Ile watches will ever be created and each will house 834 individual parts and offer 16 complications that can be read off of a double-faced watch. The Ø 47mm case houses the new Caliber 2750 with tourbillon, minute repeater, moonphase, age of moon, perpetual calendar, second time zone, equation of time, sunrise, sunset, and sky chart. The watch was created in honor of the brand's 250th anniversary, celebrated in 2005.

## ZENITH

## GRANDE CLASS TRAVELLER RÉPÉTITION MINUTES

Housing the new El Primero 4031 movement with minute repeater, alarm function (with vibrating alert or buzzer), dual time, large date and three power-reserve indicators at 6:00 (movement reserve level is shown on left side of aperture, alarm reserve on right; when the minute repeater's reserve is low, a red light comes on under the alarm indicator), this Grande Class Traveller Répétition Minutes contains 744 components and 63 jewels. Offering 50 hours of power reserve, it beats at 36,000 vibrations per hour and measures short time intervals to 1/10 of a second. The watch's 44mm platinum18K rose-gold case is water resistant to 50 meters and holds a curved sapphire crystal and sapphire crystal caseback.

# Minute Repeaters and Sonneries

Watches with audible indications are among the most fascinating timepieces of all. They are also the hardest to make and therefore the most exclusive of complications, as only a small number of watch companies or artisans offer them within their collections. Representing the peak of Haute Horology, minute repeaters embodying a blend of expertise and magic generally would appear in classic forms. For the past few years, however, minute repeaters have been experiencing a mini-revolution on both technical and aesthetic levels.

## The music of time

Audible indication systems appeared in pocket watches as early as the late 15th century, striking the time "in passing" like church bell towers. Watches in the late 17th century were equipped with the first mechanisms that could strike hours and quarters on demand—an extremely practical method of telling time at night in an era before electricity! Watches at that time were equipped with hammers striking a small bell. English watchmaker Thomas Mudge is said to have developed the minute repeater mechanism around 1750. At the end of the same century, the famous Abraham-Louis Breguet replaced the bells with gongs, or hardened steel blades coiled inside the case, and reduced the thickness of watches while achieving a purer sound.

1

2

### A cathedral on the wrist

In the 20th century, and by then miniaturized to wristwatch size, the minute repeater asserted itself as one of the most eloquent demonstrations of horological expertise. It takes impeccable competence to make this complex system of mechanical memory, with its "feeler-spindles" picking up information on the hour, quarter and minute "snails" in order to transmit them to "gathering-pallets." It takes an extremely keen ear to adjust the gongs just as one would tune a musical instrument, especially in a chiming or "cathedral" striking mechanism with three or four gongs. There are two types of audible indication: striking in passing (grand and small strike) and striking on request (repetition). Grand strike means that the watch repeats the hours every quarter, and small strikes signal either the full hours or the quarters. As for minute repeater mechanisms, they are generally equipped with two gongs, a low note for the hours, a high note for the minutes, and alternating high-low notes to mark the quarters. Philippe Dufour, one of the greatest contemporary horologers, created the first minute repeater wristwatch with both grand and small strike in 1992.

### Striking inventions

Minute repeaters are attracting renewed interest in recent years, with several innovations unveiling new territory for this complication. In 2005, Jaeger-LeCoultre launched the Master Minute Repeater Antoine LeCoultre, endowed with an original sound transmission system, in which the sound is sent outwards by the sapphire crystal to which the steel gong-heels are welded. The result is a sound of peerless power, richness and purity. Aesthetically, de GRISOGONO played an iconoclastic role with its Occhio Minute Repeater. Based on the principle of reflex cameras, the watch has an aperture-type dial: when the minute repeater is activated, the twelve ceramic shutters open to reveal the movement while it is performing its cathedral

▸▸ 1 Jaeger-LeCoultre, Master Minute Repeater Antoine LeCoultre
A striking innovation: The Master Minute Repeater Antoine LeCoultre by Jaeger-LeCoultre is endowed with an unprecedented sound-transmission system featuring steel gongs directly welded to the sapphire watch crystal.

▸▸ 2 de GRISOGONO, Occhio Ripetizione Minuti
de GRISOGONO plays the iconoclast with its Occhio Ripetizione Minuti, upon which the dial is replaced by an aperture reminiscent of reflex cameras.

▸▸ 3 Philippe Dufour, Grande Sonnerie
The Grande Sonnerie by Philippe Dufour (1992) was the first minute-repeater wristwatch with grand and small strike in passing.

3

1 *Audemars Piguet, Jules Audemars Minute Repeater Tourbillon Chronograph*
*The Jules Audemars Minute Repeater Tourbillon Chronograph by Audemars Piguet boasts an exclusive striking system ensuring an enhanced sound.*

2 *F.P. Journe, Sonnerie Souveraine*
*In developing his Sonnerie Souveraine, F.P. Journe ensured simple and secure operation of the watch, while also limiting the amount of energy required for the striking mechanism.*

3 *Zenith, Grande Class Traveller Minute Repeater El Primero*
*Zenith's Grande Class Traveller Minute Repeater El Primero's alarm is equipped with a two-tone striking or vibrate mode and striking-mechanism power-reserve indication.*

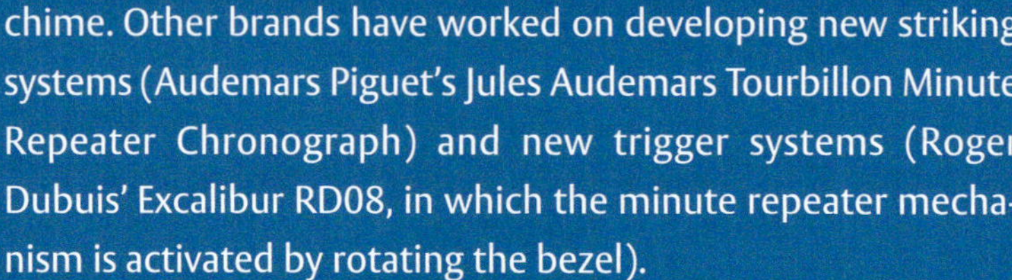

chime. Other brands have worked on developing new striking systems (Audemars Piguet's Jules Audemars Tourbillon Minute Repeater Chronograph) and new trigger systems (Roger Dubuis' Excalibur RD08, in which the minute repeater mechanism is activated by rotating the bezel).

**Auditory and visual delights**

Aside from the presence of a small sliding bolt or pushpiece along the case side on classic minute repeater and striking watches, there are no other telltale signs that indicate the presence of this sophisticated complication. Today, most watchmakers prefer to show the mechanism by opening the dials. Such is the case with the Zenith Grande Class Traveller Répétition Minutes El Primero with its triangular aperture and its openworked mainplate. Composed of 744 parts, the chronograph movement drives an alarm function and a dual time-zone display, as well as indications of two power reserves, one for the watch and the other for the striking mechanism. The same desire for openness, albeit more discreet, is revealed in the Sonnerie Souveraine by F.P. Journe, where the

1

2

ZENITH
REPETITION MINUTES
RESERVE DE MARCHE
Heures
Minutes
DT
3

▸▸ 1 *Harry Winston, Westminster Tourbillon*
*The Westminster Tourbillon by Harry Winston combines a flying tourbillon with a minute repeater that mimics the famous Big Ben chime.*

▸▸ 2 *Vacheron Constantin, Maîtres Cabinotiers Skeleton Minute Repeater*
*Record-breaking slenderness and transparency are the keynotes for the Maîtres Cabinotiers Skeleton Minute Repeater by Vacheron Constantin, comprising more than 330 openworked parts amounting to a total thickness of just 3.3mm.*

▸▸ 3 *Ulysse Nardin, Genghis Khan*
*Ulysse Nardin's Genghis Khan is equipped with automata that spring to life when the striking mechanism is activated.*

2

hammers appear through a cut-out at 10:00. For this "grand strike" model that required almost six years to develop and create, François-Paul Journe developed systems facilitating the use of the watch as well as enhancing its security (the Sonnerie Souveraine was designed, as Journe says, to be safe in the hands of an eight-year-old child) and reducing the energy required for the striking mechanism. No fewer than ten patents protect the innovations it now embodies. The Westminster Tourbillon by Harry Winston combines a flying tourbillon with a minute repeater reproducing the famous Big Ben chime. The absence of a dial enables one to admire the four hammers striking the four cathedral gongs. However, the prize for most easily observed in action must go to Vacheron Constantin's Maîtres Cabinotiers Skeleton Minute Repeater, unveiled in 2006. The movement is mere 3.3mm thick yet comprises more than 330 parts, most of them entirely openworked, decorated and chased—a real treat for the eye as well as the ear!

3

**Movin' to the music**

Ulysse Nardin's Genghis Khan model is the first minute repeater, tourbillon and jacks model equipped with a Westminster chime. Its 18-karat gold case houses four gongs striking different notes (E-C-D-G). It marks the hour with a G, the minute with an E and indicates the quarters on four gongs thanks to three different sound sequences. Genghis Khan's black onyx dial is adorned with hand-engraved gold figurines that spring to life when the striking mechanism is activated, each moving in perfect synchronization to the beat of the gongs.

## AUDEMARS PIGUET

## JULES AUDEMARS MINUTE REPEATER TOURBILLON WITH CHRONOGRAPH

## REF. 26050OR.00.D002CR.01

Cased in 18K rose gold, the Jules Audemars Minute Repeater Tourbillon with Chronograph is powered by the manual-winding Caliber 2874. The watch's minute repeater device strikes on request—and two notes—the hours, quarter hours and minutes, transcribing time that we see into time that we hear. The chronograph has a 30-minute counter and small seconds along the tourbillon axis. Limited production.

## BOVET

### THE MINUTE REPEATER WITH TOURBILLON AND REVERSED HAND FITTING

This single watch encompasses everything BOVET is about. The 44mm 18K white-gold case houses a minute repeater with tourbillon and reversed hand-fitting movement of 365 components visible through a crystal sapphire. The unique feature of this hand-wound caliber is that the movement is shown side up meaning that the hand-fitting has to be reversed so that the hands turn clockwise dial-side. The BOVET minute repeater presents 2 different cathedral chiming gongs, a low-pitch first striking the hours, followed by a combination of high- and low-pitch gongs chiming the quarters and then the minutes.

## BOVET

## THE MINUTE REPEATER

This 18K white-gold, 44mm BOVET case houses a minute repeater finished with the highest standards of Swiss watchmaking. The mother-of-pearl dial features a miniature hand-painted Siberian Tiger. The oscillating mass is engraved in fleurisanne style and the steel screws and cathedral chiming gongs blued by fire are BOVET trademarks. The BOVET minute repeater has 2 different cathedral chiming gongs. The hammers first strike the low-pitch gong to count the hours from 1 to 12, then a combination of high- and low-pitch gongs from 0 to 3 chime the quarters, and then from 0 to 14 high-pitch gongs chime the minutes after the quarter. The chiming work of the minute repeater can be seen through the sapphire crystal.

## BVLGARI

## BVLGARI-BVLGARI - REF. BBW40C6GLRM

The Bvlgari-Bvlgari Ref. BBW40C6GLRM, features a completely hand-made, mechanical, manual-winding Vaucher Caliber 9950 movement, with 31 jewels, and a vibration rate of 21,600 vph. It offers the functions of hours, minutes, seconds and minute repeater. The 40mm, polished platinum case, which is fitted with an antireflective, scratch-resistant, sapphire crystal and a snap-on back displaying the movement through a sapphire crystal, is engraved with a limited-edition number on the caseback and is water resistant to 30 meters. The blue satiné dial, embellished with Côtes de Genève and stippling, has a blue galvanic lower-level minute circle, a small seconds counter, hand-applied indexes and displays hours, minutes and seconds. The hand-sewn brown alligator strap has a platinum triple-fold-over buckle. Limited edition of 15 pieces.

## CHOPARD

## L.U.C STRIKE ONE - REF. 16/1912

The L.U.C Strike One houses the L.U.C 96 SH caliber, which measures 33mm in diameter and is 5.7mm thick, offers 65 hours of power reserve (Twin Technology—two spring-barrels). It provides an hour-strike function with on/off function and a small seconds indicator at 6:00. The movement bears the Geneva Seal quality hallmark and is chronometer-certified. The bridges are decorated with straight-line Côtes de Genève pattern. Limited edition of 100 timepieces in 18K white gold.

## DANIEL ROTH

## ELLIPSOCURVEX MINUTE REPEATER - REF. 308.Y.60.152.CN.BD

This elegant 18K white-gold timepiece features an entirely hand-decorated in-house patented manual-winding minute-repeater movement. The 18K white-gold guilloché dial displays a black enamel hour indication encircled with a white-gold ring, a tiny striking indicator aperture above and power-reserve sector below the hours and minutes hands, as well as small seconds at 7:00. This movement with two hammers, a flat balance-spring and escapement lever, is housed in an Ellipsocurvex white-gold case, the repeating pushpiece is located on the case at 9:00. The watch is fitted with a sapphire caseback and is water resistant to 3atm.

## DANIEL ROTH

## LA GRANDE SONNERIE - REF. 607.X.60.166.CN.BD

In this extraordinary timepiece, Daniel Roth has succeeded in accommodating within its famed double-ellipse case a Grande Sonnerie mechanism complete with minute repeater, grand- and small-strike system, a four-hammer Westminster chime, a tourbillon and two power-reserve indicators—one for the striking mechanism (sector between 12:00 and 3:00) and the other for the movement (sector between 3:00 and 6:00). Five patented systems protect the movement and the functions during handling.

## GERALD GENTA

## ARENA GRANDE SONNERIE - REF. GS1.X.60.529.CN.BA

The Arena Grande Sonnerie features a manual-winding in-house tourbillon Grande and Petite Sonnerie movement with minute repeater and double power reserve. This Grande Sonnerie, presented for the first time in 1994, was improved and the new version presented for the 30th anniversary of the Gérald Genta brand. The very limited space of this wristwatch's movement with off-center hour display, tourbillon and two barrels also houses the repeater mechanisms and the "au passage" strike-work with a four-hammer Westminster chime. Located on the caseback of this extraordinary piece are subdials for the power reserves (one for the Sonnerie and one for the movement), as well as for the sonnerie mode selection (grande, petite, mute).

## GERALD GENTA

## OCTO GRANDE SONNERIE - REF. OGS.Y.60.930.CN.BD

The Gérald Genta Octo Grande Sonnerie offers an entirely hand-decorated automatic Grande sonnerie tourbillon movement, with grand and small strike on four Westminster chime gongs. It also presents retrograde hours in the purest tradition of the Gérald Genta brand, as well as revolving minutes disk and minute repeater functions. It offers a power reserve of 48 hours for the movement and 18 hours for the striking mechanism, shown on two separate displays. This amazing movement is housed in a Ø 42.5mm octagonal white-gold case and enhanced by a white-gold guilloché and cloisonné black and red ceramic dial.

## GERALD GENTA

## OCTO MINUTE REPEATER - REF. ORM.Y.50.785.CN.BD

The Octo Minute Repeater features a Gérald Genta exclusive automatic manufacture movement with 37 jewels, oscillating at 21,600 vibrations per hour and offering a 42-hour power reserve. This two-gong minute repeater mechanism is combined with a jumping hour and retrograde minutes. It is housed in an octagonal 5N red-gold case with open back, sporting a unique ceramic cloisonné dial on an engraved red-gold base as well as a falcon-eye cabochon-set pearl-beaded crown. This model is presented on a hand-stitched black folded alligator strap with circular-grained gold folding clasp. It is water resistant to 3atm.

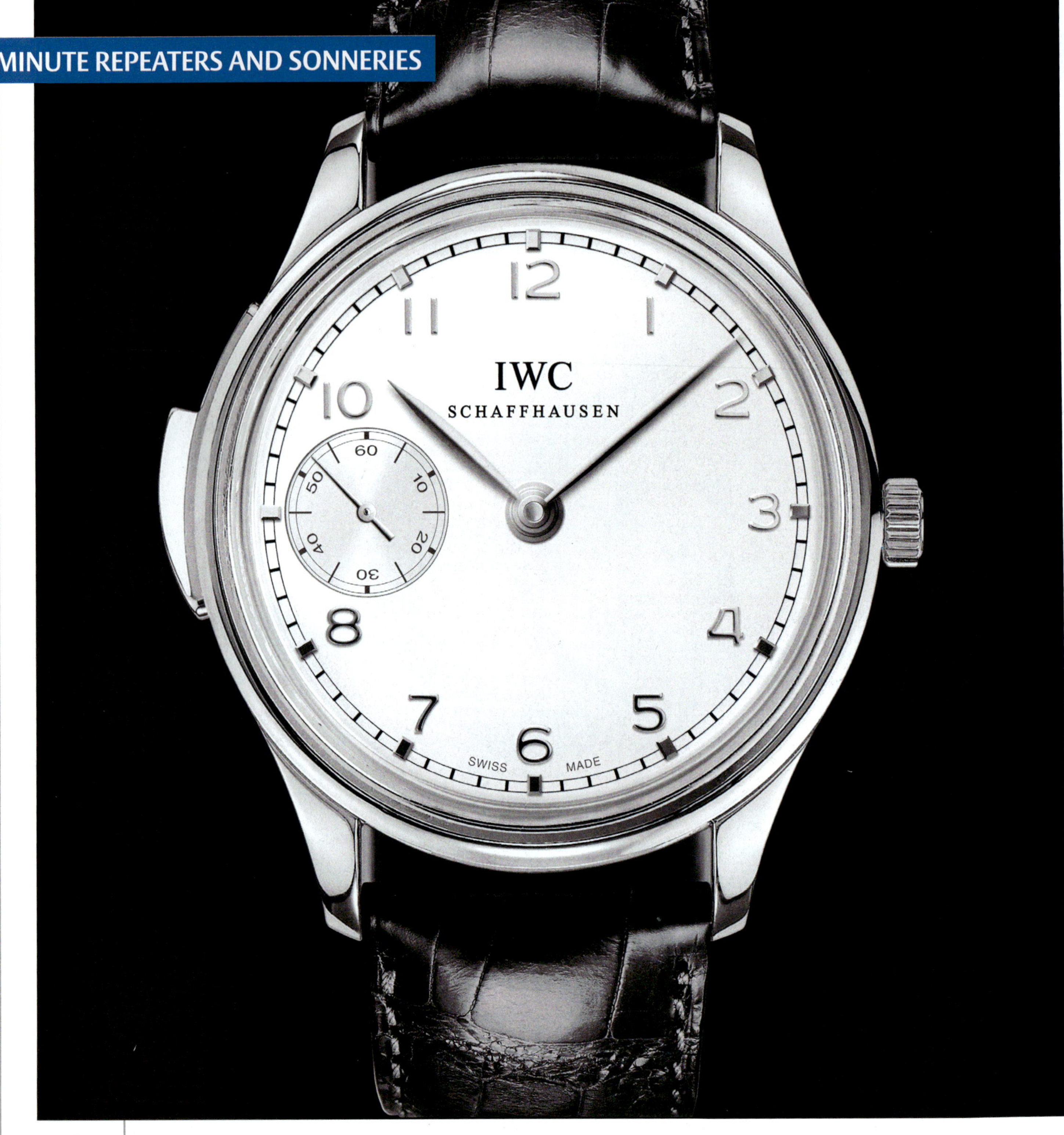


## IWC

## PORTUGUESE MINUTE REPEATER - REF. IW524204

The Portuguese Minute repeater houses the 95290 caliber with 54 rubies and beating at 18,000 vibrations per hour. It offers 43 hours of power reserve and features a stunning minute repeater function with small second and stop device. It is offered in limited editions of platinum and two colors of gold.

## JAEGER-LECOULTRE

## MASTER MINUTE REPEATER ANTOINE LECOULTRE

Crafted in titanium, the acoustical qualities of this timepiece surpass all expectations and offer a new standard for repeaters. Its crystalline chime, which has exceptional purity of sound, is the fruit of a collaborative effort between a musical conductor and the watchmakers and technicians at Jaeger-LeCoultre. The materials used for transmitting the vibrations at greater speed are new: the gongs are made of steel and sapphire crystal acts as the sound transmitter. The manually wound caliber 947 mechanism is endowed with 360 hours (15 days) of power reserve—indicated on the dial. Created in a limited edition of 200 pieces, the Master Minute Repeater Antoine LeCoultre houses 41 jewels and beats at 21,600 vibrations per hour.

## PARMIGIANI FLEURIER

## KALPA XL HEBDOMADAIRE PALLADIUM 30TH ANNIV. MODEL - REF. PF011810.01

Michel Parmigiani celebrated his 30th year in watchmaking with special models such as this extraordinary Kalpa XL. It houses the mechanical self-winding PF Caliber 118, the brand's proprietary openworked 18K solid gold movement boasting an 8-day power reserve with patented spherical differential (at 12:00), double openworked barrel, hand-beveled and -engraved bridges, steel parts drawn out with a file and 21,600 vph. The 950 palladium case measures 53x37.2mm and 11.2mm thick and is completed with a grained bezel ring, transparent sapphire crystal dial, antireflective curved sapphire crystal and caseback. Fitted on a Hermès alligator leather strap, this model is limited to 30 pieces.

## PARMIGIANI FLEURIER

## TECNICA XI SORBA

This one-of-a-kind piece was created for a client who wanted a watch inspired by nature. Michel Parmigiani based his design on the rowan tree and added a Fibonacci development, realized in the spiral. The manual-winding PF 351.03 with tourbillon, minute repeater on 2 timbers, perpetual calendar, and 43-hour power reserve powers Tecnica XI Sorba. The hand-engraved 18K gold dial has black transfers for the timer, counter and logo, rhodium-plated steel hands and blued-steel counters. The 950 platinum 42mm case is topped with a polished-finish bezel with double knurling; a 5.5mm blue sapphire serves as the crown. The double sapphire crystal caseback with hinge is hand engraved and covered with baked translucent enamel. The watch is water resistant to 30 meters and fitted on a "prestige" alligator leather strap.

## ULYSSE NARDIN

## CIRCUS MINUTE REPEATER - REF. 746-88

This alluring Circus Minute Repeater, Ref. 746-88, is crafted in 18K rose gold and houses the UN-74 movement. The Minute Repeater chimes the hours, quarter hours and minutes on two distinct gongs. The animated Jaquemarts, a specialty of Ulysse Nardin, are crafted in 18K gold and are hand-made and -applied. This watch is built in a limited edition of 30 pieces in gold and 30 pieces in platinum.

## ULYSSE NARDIN

## GENGHIS KHAN MINUTE REPEATER - REF. 780-88

Crafted in 18K white gold, the award-winning Genghis Khan Minute Repeater, Ref. 780 88, houses the UN-78 movement. It offers a Westminster Carillon to chime the time, and is equipped with a one-minute tourbillon escapement and animated Jaquemarts in 18K gold. This watch is built only in a limited edition of 30 pieces in white gold, though it is also available in 18K rose gold in a limited edition of 30 pieces.

## ULYSSE NARDIN

## HOURSTRIKER SONNERIE EN PASSANT - REF. 756-88/E2

The Hourstriker Sonnerie en Passant, Ref. 756-88/E2, is crafted in 18K rose gold in a 42mm case. It houses the UN-75 movement with 41 jewels, and features an hour striker and animated Jacquemart in 18K gold on an onyx dial. The watch has an exhibition caseback. It is also available in platinum.

## ULYSSE NARDIN

## FORGERONS MINUTE REPEATER - REF. 719-61/E2

This Forgerons Minute Repeater, Ref. 719-61/E2, is crafted in a 42mm platinum case. It houses the UN-71 manual-wind movement with 40 hours of power reserve. The exquisite timepiece features animated Jaquemarts that move with the repeater. The repeater functions on two distinct gongs. The dial of this watch is onyx. It is built in a limited edition of 50 pieces. It is also available in 18K rose gold with various dial combinations.

## VACHERON CONSTANTIN

## MALTE PERPETUAL CALENDAR MINUTE REPEATER - REF. 30040/000R

Crafted and accented in 18K pink gold, this Malte Perpetual Calendar Minute Repeater offers minute repeater and perpetual calendar with day, date, month, year, and moonphase. Powered by the extra-flat, mechanical manual-winding Vacheron Constantin Caliber 1755 QP, the model was released in a 30-piece limited edition. With a perpetual calendar module, the 30-jeweled movement measures 4.9mm thick, is finished and decorated by hand, and is visible through a sapphire caseback. The silvered dial is guilloché with a basket-weave pattern.

## VACHERON CONSTANTIN

## MINUTE REPEATER PERPETUAL CALENDAR - REF. 30020

The platinum Minute Repeater Perpetual Calendar houses the sophisticated, manually wound Caliber 1755 QP offering perpetual calendar (day, date, month, leap year and moonphase) and repeater functions (hours, quarter hours and minutes chime on demand). The 30-jeweled movement beats at 18,000 vibrations per hour and offers 34 hours of power reserve. This watch features a sapphire caseback.

## VACHERON CONSTANTIN

## PLATINUM SKELETON MINUTE REPEATER - REF. 30030/000P

Fully skeletonized, this Platinum Skeleton Minute Repeater is fitted with the manually wound, extra-flat Vacheron Constantin Caliber 1755 SQ. This exceptional 30-jeweled movement is 3.3mm thick and is entirely decorated and chased by hand. The repeater is operated via a slide on the left of the case. The Platinum Skeleton Minute Repeater was created in a limited edition of just 15 numbered pieces.

## VACHERON CONSTANTIN

## MINUTE REPEATER, REF. 30010, AND MINUTE REPEATER WITH PERPETUAL CALENDAR - REF. 30020

A proprietary Vacheron Constantin design, the minute-repeater caliber 13" features center hour and minute hands, two-position stem for winding and setting, and offers 40 hours of power reserve. Crafted in either 18-karat rose gold or platinum, the minute repeater features an open-worked decoration and strikes the hours, quarters and minutes on a pair of tiny hammers. The perpetual calendar offers hour, minute, date, day of week, month, moonphase, and takes into account leap years. The movement houses 30 jewels, and beats at a frequency of 18,000 vibrations per hour.

## ZENITH

## GRANDE CLASS TRAVELLER RÉPÉTITION MINUTES

Cased in platinum and water resistant to 50 meters, this Grande Class Traveller Répétition Minutes houses the El Primero 4031 automatic chronograph movement with minute repeater, alarm function (with vibrating alert or buzzer), dual time, large date and three power-reserve indicators at 6:00 (movement reserve level is shown on left side of aperture, alarm reserve on right; when the minute repeater's reserve is low, a red light comes on under the alarm indicator). The chronograph offers central seconds hand, 30-minute counter and 1/10-of-a-second counter while the minute repeater functions are controlled by the single pushbutton on the crown. The incredibly complex El Primero 4031 consists of 744 components, including 63 jewels. It beats at 36,000 vibrations per hour and offers over 50 hours of power reserve. The 22K white-gold oscillating weight is engraved with a Grain d'Orge guilloché pattern.

## ZENITH

## GRANDE CLASS TRAVELLER RÉPÉTITION MINUTES

The Grande Class Traveller Répétition Minutes houses the El Primero 4031 automatic chronograph movement with minute repeater, alarm function (with vibrating alert or buzzer), dual time function, large date and triple power-reserve indicators (movement, alarm and minute repeater). The chronograph offers central seconds hand, 30-minute counter and 1/10-of-a-second counter while the minute repeater functions are controlled by the single pushbutton on the crown. The incredibly complex El Primero 4031 consists of 744 components, including 63 jewels. It beats at 36,000 vibrations per hour and offers over 50 hours of power reserve. The 22K white-gold oscillating weight is engraved with a Grain d'Orge guilloché pattern. The platinum watchcase is water resistant to 50 meters.

# Tourbillons

The tourbillon is one of the pinnacles of the watchmaking art. It takes a wealth of skill and patience to build this tiny mobile carriage, often comprising 60 to 70 parts and weighing less than one gram. The tourbillon is also the most visually spectacular of complications—and watchmakers have adopted the custom of revealing it through the dial. However, the price of such popular demand means that this formerly ultra-exclusive mechanism is on the path to becoming commonplace, thus at risk of losing much of its appeal to collectors of the unusual. Nonetheless, a number of innovative technical developments such as multi-axis tourbillons and watches with several tourbillons prove that the "star complication" has every intention of keeping discerning connoisseurs in a whirl—and prices in an upward spiral!

### A highly sophisticated mechanism

In the days when watches spent much of their time in vertical positions within waistcoat pockets, the force of gravity had a negative influence on the running of a movement, particularly on the oscillations of the balance and the balance-spring, two organs that greatly determine a watch's accuracy. To avoid such irregularity, Abraham-Louis Breguet (1747-1823) enclosed the "heart" of the watch (balance, balance-spring and escapement) within a tiny mobile carriage spinning once a minute on its axis. By successively adopting all possible vertical positions, the balance enabled the variations in rate to compensate for each other. The triumph of the wristwatch in the 20th century might have spelled the end of the tourbillon, since this device loses much of its relevance when a watch moves in all directions due to natural wrist movements, and tourbillon wristwatches were in fact rare. But the renewal of mechanical watchmaking in the 1980s revived the tourbillon to the dizzying heights of horological complications and asserted it as the supreme embodiment of Swiss expertise.

1

2

## Undimmed fascination

Since the turn of the millennium, tourbillon production has soared. In 2004 alone, total production amounted to around 2,500 watches—which was more than all tourbillons made in the 175-year span between 1805 and 1980!—and more than 50 Swiss brands offered them in their collections. These are perplexing figures when one considers that only a handful of craftsmen are truly capable of making this mechanism. Watchmakers have grasped the dangers of dispelling the tourbillon's exclusive aura and are vying with one another to find ingenious technical and aesthetic innovations that will set them apart. In 2003, Chopard established its place within the circle of great watch manufactures by presenting its own tourbillon movement, the Caliber L.U.C 1.02 4T with four barrels ensuring an eight-day power reserve. Zenith has developed a tourbillon movement based on its famous El Primero chronograph caliber (see Chronographs chapter); in Zenith's Grande ChronoMaster XXT Tourbillon, the tourbillon's technical accomplishment is combined with an extremely original aesthetic appeal featuring a dial with a distinctive asymmetric charm. In the spring of 2006, Jaeger-LeCoultre caused a stir with the Master Tourbillon, powered by a proprietary movement distinguished by its refined technical features (particularly in terms of energy) as well as its price, which is far more "affordable" than competing pieces. This irked quite a few people in the watch industry and led many consumers to wonder whether the prices of some tourbillons might be overstated.

## Clear beauty

The tourbillon's mobile carriage spins on its axis like a planet, opening the doors for creative plays on composition and transparency. Girard-Perregaux set the tone with its famous Tourbillon with three gold Bridges featuring a highly original geometrical layout. Within the Royal Blue Tourbillon by Ulysse Nardin, the clear sapphire bridges and mainplate create the impression that

3

- 1 Audemars Piguet Tourbillon
  *In 1986, Audemars Piguet presented the first tourbillon wristwatch produced in a small series.*
- 2 Zenith, Grande ChronoMaster XXT Tourbillon
  *The Zenith Grande ChronoMaster XXT Tourbillon's asymmetrical dial covers a movement beating at 36,000 vibrations per hour.*
- 3 Girard-Perregaux, Tourbillon with Three Gold Bridges (skeleton version)
  *Skeleton version of the Tourbillon with Three Gold Bridges by Girard-Perregaux.*
- 4 Jaeger-LeCoultre, Master Tourbillon
  *The Master Tourbillon by Jaeger-LeCoultre was launched in 2006 and caused a sensation due its technology coupled with a well-below average price for a tourbillon.*
- 5 Chopard, L.U.C Tourbillon
  *The L.U.C Tourbillon watch by Chopard features a variable-inertia balance as well as four barrels ensuring an 8-day power reserve.*

4

5

▸▸ 1 *Vincent Calabrese, Regulus*
*Vincent Calabrese provides the ultimate in transparency with his Regulus model.*

▸▸ 2 *Ulysse Nardin, Royal Blue Tourbillon*
*The Royal Blue Tourbillon from Ulysse Nardin features sapphire bridges and a sapphire mainplate.*

▸▸ 3 *Richard Mille, RM009 Felipe Massa*
*The RM009 Felipe Massa by Richard Mille boasts the world's lightest case and movement with a combined weight of less than 30 grams.*

▸▸ 4 *Harry Winston, Opus VI*
*In the Opus VI, Greubel Forsey devised a tourbillon for Harry Winston that appears to be entirely freed from its traditional carriage.*

▸▸ 5 *Richard Mille, RM012*
*The Richard Mille RM012's gear train and flying tourbillon are incorporated within an ultra-contemporary tubular structure.*

▸▸ 6 *MB & F, Horological Machine N° 1*
*The three-dimensional design of the Horological Machine N° 1 by MB&F places the tourbillon in the center between two dials.*

▸▸ 7 *Jaeger-LeCoultre, Gyrotourbillon I*
*The Jaeger-LeCoultre Gyrotourbillon I is distinguished by its "spherical tourbillon" comprising two carriages fitted together at a 90° angle.*

1

the tourbillon is hanging in the void. However, the ultimate in transparency is offered by Vincent Calabrese in his Regulus model, equipped with a flying tourbillon movement that is totally isolated in space. Meanwhile, the lightweight champion was presented by Richard Mille with his Felipe Masse RM009: composed of avant-garde materials, the case and movement combined weigh less than 30 grams, making it the world's lightest to date. In 2006, Richard Mille shifted the contemporary and minimalist character of his calibers into an even higher gear by introducing the RM012. The entire going-train and the flying tourbillon are completely integrated within a tubular structure that serves the functions normally handled by the mainplate. Meanwhile, the Horological Machine No 1 from MB & F (Max Büsser & Friends) gives the tourbillon a central place within an unconventional three-dimensional design featuring two separate dials for the hours and minutes, as well as four barrels supplying a seven-day power reserve.

2

3

5

4

6

7

### Above and beyond

The past few years have borne a new generation of multi-axis tourbillons, the purpose of which is to optimize the precision of wristwatches by multiplying the various spatial positions of the balance and spring and thus extending compensation for variations in rate beyond the merely vertical plane. Launched in 2004, the Gyrotourbillon I by Jaeger-LeCoultre combines a "running equation" (see the Equation of Time chapter) with a "spherical tourbillon" comprising two carriages one inside the other: the outer carriage revolves once in 60 seconds, whereas the inner carriage, inclined at a 90° angle in relation to the first, spins 2.5 times per minute. A similar principle inspired the Opus VI by Harry Winston, designed by watchmakers Robert Greubel and Stephen Forsey. The 30° angle between the rotation axes guarantees excellent chronometric performances. Opus VI's gear-trains are off-set beneath the bridge and conceal the right-hand side of the dial, thereby making the tourbillon appear to be free of its traditional carriage: detached, feather-light and the center of attention. Independent watchmaker Thomas Prescher surprised the

1

2

industry when he alone created and released his Triple Axis Tourbillon with constant force in less than one year.

### Head-spinning inventions

Jean Dunand's Tourbillon Orbital offers a double-rotation system in which the flying tourbillon spins on its axis in one minute, while performing a complete rotation around the dial in one hour. The same double revolution appears on the Polo Tourbillon Relatif by Piaget, one of the major surprises of 2006. Suspended from the tip of the minute hand, the mobile carriage of the flying tourbillon appears to be disconnected from the movement that drives it, representing a fine blend of mechanical ingenuity and visual magic. For the Blu Majesty Orbiting Tourbillon, Bernhard Lederer, the watchmaker behind the Blu brand, invented a new way of reading time by combining three axes and three speeds. The tourbillon rotates on its axis in one minute; it is housed within a carriage that spins on its own axis once one hour and serves as a minute pointer; and this carriage itself turns around the entire dial once in twelve hours, while its position indicates the hours. Lederer's masterpiece is a non-stop choreographic performance.

3

### One, two, three tourbillons

Another technical and aesthetic avenue currently being explored is that of offering two or more tourbillons on a single watch. In 2006, Breguet reinterpreted its favorite complication by introducing a Classique Grand Complication with Twin Rotating Tourbillons, duly protected by several patents. The two independent tourbillons, coupled with a differential gearing, are fixed to a plate that completes one rotation in twelve hours. They thereby mutually compensate for their infinitely small variations in rate in order to offer superlative precision. For his 3Volution, Antoine Preziuso devised a roller system spinning once every two minutes and fifteen seconds, carrying three tourbillons each performing a complete revolution about their axis in one minute. Meanwhile,

4

5

6

7

Greubel Forsey is working on a Quadruple Tourbillon à Différentiel, which should be unveiled in 2008. When can we expect to see a watch with eight, twelve or seventeen tourbillons?

## Ladies take a spin

Tourbillons appeal not only to men. In 2005, Zenith presented the first resolutely feminine tourbillon. A concentrated blend of haute horology and jewelry, the white-gold Star Tourbillon El Primero features a diamond-studded star-shaped tourbillon bridge; the watch is set with more than 280 diamonds and the mother-of-pearl dial is topped with large dancing numerals. The Kalpa XL Tourbillon Diamonds by Parmigiani Fleurier associates a one-week power reserve with a carriage performing two rotations per minute (a 30-second tourbillon) resulting in optimized rating performances on a white mother-of-pearl dial nestled within a diamond-set case. The Jules Audemars Ladies' Tourbillon from Audemars Piguet also marries mother-of-pearl and diamonds while highlighting the tourbillon through a large opening at 6:00.

8

▸▸ 1 *Thomas Prescher, Triple-Axis Toubillon*
*Thomas Prescher presented a new triple-axis tourbillon in 2005.*

▸▸ 2 *Antoine Preziuso, 3Volution*
*The 3Volution by Antoine Preziuso features three tourbillons mounted on a rotating roller.*

▸▸ 3 *Jean Dunand, Tourbillon Orbital*
*Within the Tourbillon Orbital by Jean Dunand, the tourbillon spins once on its axis in one minute, while making a complete rotation around the dial in one hour.*

▸▸ 4 *Breguet, Double tourbillon tournant*
*Breguet's Double tourbillon tournant: The two independent tourbillons mutually compensate for their tiny variations in rate.*

▸▸ 5 *Piaget, Polo Tourbillon Relatif*
*Impressive aerobatics are present in Piaget's Polo Tourbillon Relatif, where the mobile tourbillon carriage is in fact suspended from the tip of the minute hand.*

▸▸ 6 *Blu, Blu Majesty Orbiting Tourbillon*
*The Blu Majesty Orbiting Tourbillon devised by Bernhard Lederer invented a new way of reading off the time.*

▸▸ 7 *Audemars Piguet, Jules Audemars Ladies' Tourbillon*
*A shimmering mother-of-pearl and diamond mantle clothes the Jules Audemars Ladies' Tourbillon by Audemars Piguet.*

▸▸ 8 *Parmigiani Fleurier, Kalpa XL Tourbillon Diamands*
*Parmigiani Fleurier offers a feminine interpretation of the tourbillon in the Kalpa XL Tourbillon Diamants.*

▸▸ 9 *Zenith, Star Tourbillon El Primero*
*Zenith's Star Tourbillon El Primero is a spectacular encounter between Haute Horlogerie and Haute Joailerie.*

9

## AUDEMARS PIGUET

## EDWARD PIGUET SKELETON TOURBILLON - REF. 25947OR.00.D002CR.01

Created in 18K pink gold, this Edward Piguet Skeleton Tourbillon houses the manual-winding Caliber 2881 SQ. The elegant timepiece is equipped with a one-minute tourbillon with small seconds along the tourbillon axis. This complication features an elaborate hand-hollowed and -engraved rose-gold finished movement. The watch features a sapphire caseback revealing the exquisite movement.

## AUDEMARS PIGUET

## EDWARD PIGUET TOURBILLON WITH LARGE DATE DISPLAY

## REF. 26009BC.OO.D002CR.01

This 18K white-gold Edward Piguet Tourbillon with Large Date Display features a manual-winding movement with tourbillon, hours, minutes, large date, and small seconds along the tourbillon axis. The dial features an engine-turned pattern with applied riveted numerals and hands in gold. This watch's case is glare-proof and has a sapphire crystal and caseback. Limited production.

## AUDEMARS PIGUET

## JULES AUDEMARS 10-MINUTE REPEATER WITH TOURBILLON

## REF. 26072TI.00.D002CR.01

Crafted in titanium, this Jules Audemars features a minute repeater device that strikes on request—on two notes—the hours, 10 minutes, and minutes, with small seconds along the tourbillon. The dial is engine-turned with a sun pattern and riveted gold numerals with gold hands. The timepiece is equipped with an antireflective sapphire crystal and antireflective sapphire crystal caseback. Also available in rose gold.

## AUDEMARS PIGUET

## LADY JULES AUDEMARS TOURBILLON - REF. 26084OR.ZZ.D016CR.01

This Lady Jules Audemars Tourbillon houses a manual-winding rhodium-plated movement with Côtes de Genève decorative pattern and circular graining. The movement's bridges are hand-engraved with a "dancing spirals" motif. This 18K pink-gold Jules Audemars has a white mother-of-pearl dial engraved with a spiral flinqué design, polished Roman numerals, diamond hour-markers and polished 18K white-gold leaf-shaped hands. Its bezel holds 56 brilliant-cut diamonds, and the caseback is made of antireflective sapphire crystal that reveals the entire mechanism. Supplied with gemstone certification.

## BLANCPAIN

## LEMAN TOURBILLON SEMAINIER

Mechanical complexity, user-friendliness and useful information are three criteria fully met by the new Tourbillon Semainier. In addition to showing the hours, minutes, seconds, power reserve, and majestic flying tourbillon placed at 12:00, this newcomer also displays the day of the week, date, and week of the year. This Tourbillon Semainier beats to the rhythm of the new caliber 3725G within a 40mm diameter for enhanced legibility, crafted in 950 platinum with a black dial or in 18K red gold with an opaline dial.

## BLANCPAIN

## TOURBILLON GRAND DATE

The Tourbillon Grand Date is housed in an 18K rose- or white-gold Léman case. The self-winding watch features a tourbillon aperture at 12:00 and a grand date indicator at 6:00. The sapphire caseback displays the hand-engraved movement and 168 hours of power reserve. The movement, complete with tourbillon escapement, houses 307 individual parts and 35 rubies. The Tourbillon Grand Date is 38mm in diameter, features a sapphire caseback and is water resistant to 100 meters. Only 50 pieces are being created.

## BLANCPAIN

## TOURBILLON TRANSPARENCE

This tourbillon is a visual delight thanks to the sapphire crystal that allows viewing of the entire movement through the watch face. Created in a limited edition of 50 pieces, the watch houses the Blancpain 6925A caliber, a mechanical self-winding movement with seven days of power reserve. Comprising 307 parts, the movement offers flying tourbillon, large date display and power-reserve indicator.

## BOVET

## FLEURIER BUTTERFLY TOURBILLON

The tourbillon embodies all the ingenuity, precision and beauty of fine horology. This magnificent 42mm 18K white-gold timepiece reveals the astonishing workmanship of the original tourbillon construction—wolf's teeth and curved spokes on every wheel, serpentine jumper springs in blued steel, blue spinels, and the entire surface of the movement's 18K gold structure hand-engraved in fleurisanne style with raised engravings on both sides. The hand-wound haute horlogerie movement BOVET caliber 12BM05 has a power reserve of 110 hours. This jewelry version features bridges set with 88 baguette diamonds for 0.64 carat and an hour circle set with 60 diamonds for 0.77 carat.

## BOVET

## THE SELF-WINDING 8-DAY TOURBILLON

The robust, double shock-resistant flying tourbillon chosen for this 18K white-gold 42mm timepiece is ideally suited to wristwatches. The top of the tourbillon cage is decorated with BOVET's trademark, the stylized lotus flower. On each side of the dial, two plates engraved in fleurisanne style, frame the tourbillon cage and power reserve. To improve precision and stability, the BOVET tourbillon beats at 21,600 vibrations per hour while the 192 hours of power reserve in the fully wound springs of the twin barrels are shown by a serpentine hand. They ensure constant force to the tourbillon even if the watch isn't worn every day. Another version is available with baguette-cut diamonds totaling 7.02 carats.)

## BOVET

## FLEURIER SELF-WINDING 8-DAY TOURBILLON WITH ENGRAVED DIAL

BOVET displays extraordinary craftsmanship in the hand-engraved dial of its 42mm 18K rose-gold self-winding double shock-resistant flying tourbillon with an 8-day power-reserve. Decorated with fleurisanne engraving—a technique particular to BOVET—each dial is a unique piece of art with raised and overlapping scrolls gleaming against a stippled background. The dial, visible through a glass of rock crystal, is framed by a row of 88 round-cut diamonds. A serpentine hand shows the 192 hours of power reserve in the twin barrels' fully wound springs. The movement's bridges are also engraved in fleurisanne style.

## BREGUET

## CLASSIQUE GRANDE COMPLICATION – REF. 3355PT/00/986

This Classique Grande Complication openworked wristwatch is crafted in platinum and includes a tourbillon. The hand-wound movement is engraved by hand and features a small seconds on the tourbillon shaft and is enhanced by a compensating balance-spring with Breguet overcoil. The hour and seconds chapter rings are made of engine-turned and silvered gold. The movement is protected by a water-resistant case with sapphire crystal caseback.

## BREGUET

## CLASSIQUE GRANDE COMPLICATION - REF. 5359BB/6B/9V6 DD00

This Classique Grande Complication is crafted in 18K white gold and houses a hand-wound movement featuring a tourbillon with small seconds on the tourbillon shaft and a compensating balance-spring with Breguet overcoil. Its silvered gold dial is hand-engraved on a rose-engine and paved with 356 diamonds (approximately 0.5 carat). The bezel, lugs and caseband are paved with 134 baguettes (approximately 10.37 carats) and the case is water resistant.

## BREGUET

## CLASSIQUE GRANDE COMPLICATION - REF. 5317PT/12/9V6

This Classique Grande Complication wristwatch is crafted in platinum or 18K yellow gold and houses a self-winding movement engraved by hand, featuring a tourbillon with a small seconds on the tourbillon shaft, a 5-day power-reserve indicator and compensating balance-spring with Breguet overcoil. Its silvered gold dial is hand-engraved on a rose-engine and the water-resistant case is enhanced with a sapphire crystal caseback.

## BVLGARI

## BVLGARI-BVLGARI TOURBILLON - REF. BB40C6PLTB

The Bvlgari-Bvlgari Tourbillon features an in-house manufactured, mechanical, manual-winding Caliber 052 with tourbillon device; the movement offers a 64-hour power reserve, contains 20 jewels, has a vibration rate of 21,600 vph (3Hz) and is finished and decorated with Côtes de Genève and satiné soleil treatments. The Ø 40mm case is crafted in polished platinum and is fitted with a double-sided, antireflective, scratch-resistant, sapphire crystal, a bezel with engraved logo and a snap-on back, displaying the movement through a sapphire crystal. The limited edition number is engraved on the side of the case, which is water resistant to 30 meters. The silver dial is decorated with guilloché and satiné soleil treatments, blue applied minute track and numeral 12, and hand-applied, rhodium-plated indexes. The tourbillon is visible at 6:00 and has a blue bridge. The display includes off-center hours and minutes, a small seconds at 6:00 on the tourbillon cage and a power-reserve indicator on the back. The strap is made of hand-sewn black alligator and has a platinum triple-fold-over buckle. Limited edition of 20 pieces.

## BVLGARI

## BVLGARI-BVLGARI TOURBILLON - REF. BB38GLTB

Shown here in yellow gold, the Bvlgari-Bvlgari Tourbillon is powered by the proprietary, manual-winding Caliber 052 with 64-hour power reserve (indicated on back). The small seconds display is integrated with the tourbillon aperture at 6:00. Water resistant to 30 meters, this Ø 38mm case is enclosed with an antireflective scratch-resistant sapphire crystal and a snap-on sapphire crystal caseback that reveals the Côtes de Genève and satiné soleil treated movement. The silver, guilloché dial bears applied gold-plated indexes and is rimmed with an engraved bezel. Also available in white gold with an anthracite dial on a black alligator strap, the yellow-gold version (on a brown hand-sewn alligator strap with yellow-gold triple-fold-over clasp) is limited to 25 pieces.

## BVLGARI

## RETTANGOLO - REF. RT49PLTB

The Rettangolo features a mechanical, manual-winding Claret Skeleton with tourbillon device, 100-hour power reserve, 19 jewels and a vibration rate of 21,600 vph. The polished platinum 49mm case is fitted with an antireflective, scratch-resistant, sapphire crystal, a snap-on back, displaying the movement through a sapphire crystal, shows the limited-edition number engraved on the case side and is water resistant to 30 meters. The display shows hours and minutes, with small seconds at 6:00. The hand-sewn brown alligator strap has a platinum triple-fold-over buckle. Limited edition of 20 pieces.

## CHANEL

## J12 TOURBILLON FULL PAVE RUBIES

Created in a 38mm case of high-tech* ceramic and white gold, the J12 Tourbillon houses the manual-wind CHANEL movement 05-T.1. Beating at a rate of 21,600 vibrations per hour, it is equipped with 17 rubies. It offers 100 hours of power reserve and is water resistant to 50 meters. The single piece is adorned with 568 baguette-cut rubies (38 carats).

* High technology

## CHANEL

## J12 TOURBILLON FULL PAVE RUBIES - CASE BACK

A world premiere, both aesthetically and technically: a black ceramic bottom plate.

## CHOPARD

## L.U.C TOURBILLON STEEL WINGS CLASSIC - REF. 16/1906

The L.U.C 4T Quattro Tourbillon houses the L.U.C 1.02 caliber, which measures 29.1mm in diameter and is 6.1mm thick, offers 9 days of power reserve and features the tourbillon escapement, with aperture at 6:00. The carriage completes one full revolution around its axis once per minute. The 222-part L.U.C 1.02 caliber includes the variable inertia four-arm Variner-type balance with integrated inertia blocks. The movement features 33 jewels and beats at a frequency of 28,800 vibrations per hour. The exceptional movement is certified by the Geneva Seal hallmark. Limited series of 100 timepieces in white gold.

## DANIEL ROTH

## TOURBILLON 8-DAY POWER RESERVE - REF. 197.X.40.223.CC.BA

The new Tourbillon 8-Day Power Reserve features an in-house manufacture double-face manual-winding tourbillon movement with high-end finishing (stippled main plate, Côtes de Genève, beveled, hand-drawn and smoothed flanks and surfaces). This movement is housed in a flip-over pink-gold case with upper face presenting off-centered hours and minutes, extra-large tourbillon aperture with gold guilloché Clous de Paris background, traditional tourbillon with small seconds on its axis moving over triple-level segment. The lower face shows the 200-hour power reserve and date. This timepiece offers the possibility of personalized engraving on the caseback lid.

## DANIEL ROTH

## TOURBILLON RETROGRADE DATE - REF. 196.X.70.362.CM.BA

The Tourbillon Retrograde Date offers an in-house manufacture double-face manual-winding tourbillon movement with high-end finishing (stippled main plate, Côtes de Genève, beveled, hand-drawn and smoothed flanks and surfaces). This movement is housed in a platinum case with the 64-hour power reserve visible on the caseback, while the dial presents off-centered hours and minutes and retrograde date display, as well as a tourbillon carriage opening with small seconds on tourbillon axis moving over a dedicated ring. This model is water resistant to 3atm.

## F.P. JOURNE

## TOURBILLON SOUVERAIN WITH INDEPENDENT SECONDS

This platinum world-exclusive tourbillon timepiece is fitted with the manually wound F.P. Journe caliber FPJ1403 with tourbillon, power-reserve indicator, hours, minutes and an independent seconds subdial. It is equipped with a remonitoir, or constant force device, and is now enriched with the independent, or deadbeat seconds dial. In a deadbeat seconds watch, the seconds hand remains motionless for as long as the second has not actually elapsed. The hand only indicates the second once it has passed. In the Tourbillon Souverain with Independent Seconds, the watch is equipped with a natural deadbeat seconds device mounted on one of the wheels of the constant-force device. The movement beats at 21,600 vibrations per hour with a dedicated 4-arm balance with inertia adjustment.

## GERALD GENTA

## ARENA TOURBILLON RETROGRADE HOURS - REF. ATR.X.75.860.CN.BD

From the Arena collection, this 41mm watch is offered in platinum case with fluted caseband and palladium bezel, enhanced by a black and satin-finish metallic net-like dial displaying the tourbillon cage at 6:00, fixed with a sapphire bridge. It is equipped with an entirely hand-decorated in-house automatic tourbillon movement with retrograde hours, central minutes and seconds on the tourbillon axis. This model, offering a 64-hour power reserve, is water resistant to 10atm.

## GERALD GENTA

## OCTO TOURBILLON BLACK SPIRIT - REF. OTR.Y.76.999.CN.BD.S61

The Octo Tourbillon Black Spirit offers an exclusive Gérald Genta in-house tourbillon movement, entirely hand-decorated, black gold finish, featuring retrograde hours, central minutes hand and a transparent sapphire tourbillon bridge. It is housed in an octagonal Ø 42.5mm platinum case with tantalum bezel displaying the hours. The white-gold open dial is set with 22 baguette-cut black sapphires (1 carat) and 14 baguette-cut rubies (0.51 carat). This daring model is presented on a black folded alligator strap with grained folding clasp stamped with the brand symbol and is water resistant to 10atm.

## GERALD GENTA

## OCTO TOURBILLON INCONTRO - REF. OTL.Y.76.940.CN.BD

The Octo Tourbillon Incontro features the daring and exclusive combination of an in-house manufactured, automatic in-line lever tourbillon movement, offering a 64-hour power reserve on one side and a digital movement on the other. The tourbillon face presents a red, black and ivory ceramic dial on a white-gold base featuring retrograde hours, central minutes hand and a transparent sapphire tourbillon bridge; the openwork caseback displays the digital functions, 24-hour time indication, date, day (in three languages), second time-zone indication, chronograph and alarm. The Ø 42.5mm, octagonal platinum case with tantalum bezel, finely engraved on its back and side, is enhanced by two cabochon-set pearl-beaded crowns. This model is water resistant to 10atm.

## GERALD GENTA

## OCTO TOURBILLON RETROGRADE HOURS - REF. OTR.Y.50.930.CN.BD

The Octo Tourbillon Retrograde hours offers an exclusive Gérald Genta in-house tourbillon movement, entirely hand-decorated, featuring retrograde hours, central minutes hand and a transparent sapphire tourbillon bridge. It is housed in an octagonal Ø 42.5mm 5N red-gold case, enhanced by a 5N red-gold guilloché and cloisonné, black and red ceramic dial. This model offers a 64-hour power reserve and is water resistant to 10atm. It is also available in a platinum case with tantalum bezel and a white-gold guilloché black and red ceramic dial.

## GIRARD-PERREGAUX

## CAT'S EYE TOURBILLON - REF. 99490B53E791

Crafted in 18K white gold, this Cat's Eye Tourbillon houses the mechanical manual-winding GP 9702 caliber with a tourbillon device on a white-gold bridge. With 75 hours of power reserve, the movement holds 20 jewels and beats at 21,600 vibrations per hour. It is beveled and decorated with a Côtes de Genève design, entirely finished and decorated in-house. The bezel and lugs are set with 84 square diamonds, and the crown features a briolette diamond. The natural mother-of-pearl dial is fitted on a satin strap with a white-gold clasp bearing 16 diamonds.

## GIRARD-PERREGAUX

## SPORT CLASSIQUE LAUREATO EVO$^3$ TOURBILLON WITH THREE SAPPHIRE BRIDGES - REF. 99071

This exquisite Sport Classique Laureato Evo$^3$ Tourbillon with three sapphire Bridges houses the mechanical automatic-winding GP 9600.0S caliber with 30 jewels. The tourbillon is mounted on three sapphire bridges and the movement is equipped with a patented platinum micro-rotor under the barrel. Beating at 21,600 vibrations per hour, the watch offers 48 hours of power reserve. The three-piece 42mm case is crafted of titanium and platinum with a curved sapphire crystal and sapphire caseback. The watch is entirely finished by hand in-house and is water resistant to 10atm.

## GREUBEL FORSEY

## FIRST INVENTION: DOUBLE TOURBILLON 30° - EXPERIMENTAL WATCH®

The Double Tourbillon 30° houses a patented hand-wound movement consisting of two tourbillon carriages. The Caliber GF 02 features an exterior tourbillon cage with a rotating cycle of four minutes, and an interior tourbillon inclined at 30 degrees with a rotating cycle of one minute. The complete movement has 301 components and 39 jewels. The watch offers power reserve of 72 hours, and is equipped with twin rapid-rotating barrels—one of which is equipped with a slipping spring. The movement beats at 21,600 vibrations per hour and features plates and bridges of frosted nickel silver finished in jade gold. The Double Tourbillon 30° is offered in 18K yellow, white or rose gold, and platinum. The platinum-cased version features a stunning gold and mother-of-pearl dial.

## GREUBEL FORSEY

## SECOND INVENTION: QUADRUPLE TOURBILLON À DIFFÉRENTIEL SPHÉRIQUE – EXPERIMENTAL WATCH®

The Quadruple Tourbillon à Différentiel Sphérique Experimental Watch® features four tourbillon escapements (carriages). The two separate inclined regulating organs are linked with a Greubel Forsey spherical differential. This micro-mechanism enables optimum averaging out of the slightest variations in rate due to gravitational pull. These two double carriages are housed asymmetrically in the movement. The mechanism is visible through the dial and caseback as well as partially through a sapphire crystal on the side of the case. The inner tourbillons rotate once per minute and the exterior tourbillons rotate once every four minutes. The watch features a power-reserve indicator and small-seconds readout via a double-ended hand. The 43.5mm watch is a fundamental invention that is still in the experimental phase and delivery is not expected until 2008.

## GUY ELLIA

## TIME SQUARE 2311 TOURBILLON MAGISTERE

Designer Guy Ellia loves a challenge. For this reason, he has designed a skeletonized, mysterious-winding tourbillon. The Time Square 2311 Tourbillon Magistere has been created for Guy Ellia by Swiss manufacturer Christophe Claret, whose Caliber GE-R97 features 110 hours of power reserve, mysterious winding, one-minute tourbillon, skeleton barrel and ratchet-wheel. The tourbillon cage is entirely hand chamfered and has 18K gold bridges. The watch is offered in 18K white gold or platinum.

## GUY ELLIA

## TIME SQUARE 2311 TOURBILLON MAGISTERE

This version of the Time Square Tourbillon Magistere is also powered the Christophe Claret-designed Calibre GE-R97 with 110 hours of power reserve, mysterious winding, one-minute tourbillon, skeleton barrel and ratchet-wheel. The movement beats at 21,600 vibrations per hour and the tourbillon cage is entirely hand chamfered and features bridges in 18K gold. The watch is offered in platinum or 18K pink or white gold. This model's 18K white-gold case is set with 535 brilliants weighing 8.25 carats.

## GUY ELLIA

## TIME SQUARE 2311 TOURBILLON MAGISTERE

This platinum version of the Time Square Tourbillon Magistere is also powered the Christophe Claret-designed Calibre GE-R97 with 110 hours of power reserve, mysterious winding, one-minute tourbillon, skeleton barrel and ratchet-wheel. The movement beats at 21,600 vibrations per hour and the tourbillon cage is entirely hand chamfered and features bridges in 18K gold. The watch is also offered in 18K pink or white gold.

## HD3

## VULCANIA

Inspired by the world of Jules Verne, Vulcania is Fabrice Gonet's follow-up to the Raptor tourbillon. Developed with BNB, Vulcania's exceptional gyrotourbillon movement rotates once per minute on its first axis and once every 30 seconds on the second axis. All of the watch's details adhere to a nautical theme: hours are shown on a wheel through a lateral porthole at 9:00; minutes are displayed on a disk like a ship's Chadburn telegraph (360°); the120-hour power-reserve indicator is reminiscent of a sextant; and there is a porthole through which the winding mechanism of the movement can be seen. On the sapphire crystal caseback, a porthole on the main plate is similar to those of Verne's Nautilus submarine and a map of the mysterious Vulcania Island and its coordinates are engraved. Limited edition of 11 pieces in platinum and titanium.

## HARRY WINSTON

## EXCENTER TOURBILLON

This world-premiere Excenter timepiece, developed with Swiss-based English master watchmaker Peter Speake-Marin, is endowed with a double power-reserve indicator totalling 110 hours. Because this retrograde double indicator is on the rear bridges of the movement, the watch dial features a warning at 6:00. Engraved on the sapphire crystal, the silvered HW logo turns electric blue 25 hours prior to complete unwinding. The hand-winding HW400A movement beats at 28,800 vibrations per hour, and is equipped with blue sapphires rather than the traditional rubies. While all the other bridges are adorned with entirely hand-bevelled blued-steel screws, polished white screws embellish the tourbillon bridge. The dial has a ruthenium finish with 18K white-gold hour and minute circles. Water resistant to 30 meters, the Excenter Tourbillon is offered in platinum, 18K rose gold, and in gem-set versions.

## HARRY WINSTON

## OPUS VI

The Harry Winston Opus VI is a 30-degree double tourbillon developed in conjunction with Greubel Forsey. The advantage of the 30-degree angle is that the balance can oscillate constantly in all planes to achieve a more precise timing. Hours and minutes are at 3:00 and seconds are at 11:00, but the Opus VI's movement is the piece's primary focal point. Its bridge is screwed to the movement's mainplate and a balance wheel moves freely above the mainplate, thus liberating the double tourbillon of any superstructure and providing an unobstructed view of the mechanism—gone are the gear trains that generally obscure visual access. The hand-winding tourbillon mechanism offers 72 hours of power reserve.

## HARRY WINSTON

## THE PROJECT Z3 TOURBILLON SPORT SPIRIT

The Project Z3 Tourbillon Sport Spirit version features a rubber strap and a charcoal black dial with SuperLumiNova blue hands and numbers. Water resistant to 100 meters, the watch is crafted in Zalium and houses the mechanical self-winding HW401A tourbillon movement, which beats at 28,800 vibrations per hour. The movement contains an internal carriage rotating system and offers 110 hours of power reserve with a low-power-reserve warning display. The sapphire caseback reveals the movement with its Z-shaped oscillating weight. Only 50 pieces will be made.

## HARRY WINSTON

## THE PROJECT Z3 TOURBILLON VINTAGE SPIRIT

The Project Z3 Tourbillon Vintage Spirit houses an automatic tourbillon movement with hour/minute display offset at 12:00. Developed by Peter Speake-Marin, the watch features a decorative motif on the steel-colored dial that is directly inspired by the dashboards of vintage English cars. The mechanical self-winding tourbillon movement, HW401A, beats at 28,800 vibrations per hour and features an internal carriage rotating system. It offers 110 hours of power reserve with a low-power-reserve warning display. The 44mm watch is cased in Zalium and fitted on a gray patent alligator strap.

## HARRY WINSTON
## TOURBILLON GLISSIERE

The self-winding Tourbillon Glissiere employs a lineage system with two sliding blocks of platinum with toothed racks that engage the winding ratchet. Designed in a locomotive style, this watch features a steam-age power-reserve indicator along the bottom in a linear configuration and the movement is visible through lateral windows in the upper and lower case. This openworked movement beats at 21,600 vibrations per hour and offers flying tourbillon and 120-hour power reserve. Water resistant to 30 meters, Tourbillon Glissiere is produced in 25 white-gold pieces and 25 rose-gold pieces. A one-of-a-kind piece has been crafted in white gold and set with 216 brilliants and 180 baguette diamonds totaling 17 carats.

## HUBLOT

## BIGGER BANG

Cased in 18K red gold, this Bigger Bang's red-gold bezel is fastened by Hublot's H-shaped screws. Powered by the new HUB 1400 CT movement, the 44.5mm watch has a rubber-topped crown, is fitted on an integrated textured rubber strap, and is water resistant to 50 meters. The openwork dial reveals the chronograph driven from the 1-minute rotating tourbillon carriage, which is raised 2.8mm for maximum visibility.

## HUBLOT

## BIGGER BANG

This 44.5mm tantalum-cased model features the Bigger Bang's red pushpiece that both resets the chronograph to zero and causes the mechanism to stop, due to the rotary movement of the coupling wheel engaging with the Tourbillon system. Its exceptional HUB 1400 CT movement is visible through the antireflective sapphire crystal caseback and the watch is available on Hublot's integrated textured rubber strap.

## HUBLOT

## BIGGER BANG

Limited to 18 pieces, this 44.5mm Bigger Bang is housed in a black, polished ceramic case topped with a black, brushed ceramic bezel attached with Hublot's H-shaped screws. Like all Bigger Bangs, this version is fitted on an integrated rubber strap, is water resistant to 50 meters, and is powered by the tourbillon column-wheel chronograph HUB 1400CT movement (visible through an antireflective sapphire crystal caseback). A 3-piece limited edition is available in black ceramic featuring a bezel set with 48 white or black baguette diamonds totaling 2.8 carats.

## HUBLOT

## HUB 1400 CT MOVEMENT

Representing a powerful demonstration of know-how and cutting-edge technology, the Bigger Bang movement is composed of 269 components. This new Hublot HUB 1400 CT tourbillon movement beats at a frequency of 21,600 vibrations per hour (3 Hz) and is endowed with a 120-hour power reserve. The mainplate is circular-grained, decorated and topped by a unique bridgework in the shape of two stylized watch wheels, creating the illusion of being part of the wheel train they hold.

## IWC

## PORTUGUESE TOURBILLON MYSTERE - REF. IW504207

Housed in the Portuguese case, the Tourbillon Mystere is powered by the IWC 50900 caliber with 7-day power reserve. The watch features an elevated tourbillon that makes it appear to move freely and entirely independent of any other drive-elements. This watch is a limited edition of 250 watches in 18K white gold. It offers a minute tourbillon, Pellaton automatic-winding mechanism, power-reserve display and small seconds at 9:00.

## JEAN DUNAND

## PIECE UNIQUE TOURBILLON ORBITAL

The patented Tourbillon Orbital is the first wristwatch of its kind to feature a one-minute flying tourbillon that orbits the dial once an hour on a revolving movement. The watch houses the IO 200 movement designed by Christophe Claret incorporating a series of wheels within wheels. The barrel and the flying tourbillon, set opposite each other, orbit the center between two plates held apart by pillars rotating on ball bearings. The top plate consists of the revolving dial (with an aperture revealing the raised tourbillon). The barrel unwinds against a central fixed pinion, driving itself and the tourbillon around. The tourbillon cage rotates once a minute against the fixed circumference wheel through a train of wheels and regulates the speed of rotation. The wheel train reduces the necessary number of jewels in the movement to 14, thereby also reducing friction.

## JEAN DUNAND
## TOURBILLON ORBITAL

It took Christophe Claret two full years to work out the mechanical solution to winding and setting a rotating movement with a mainspring barrel that never stays in one place. The solution is vertical winding and setting conducted through a folding key that is set into the caseback. Lifting the key ring engages a central wheel that turns the ratchet wheel to wind the barrel spring. Puling out the key connects the motion work of the hours and minutes hands to set them. The watch features the power-reserve gauge on the case side rather than on the dial—another world's first for which Jean Dunand has been granted a patent.

## JACOB & CO.

## QUENTTIN

This one-minute tourbillon is powered by a vertical mechanical movement and manual-winding escapement with 31-day power reserve. Seven barrels house multiple mainsprings that interact to drive the chain, generating constant power and torque to achieve such a high power reserve. Hours, minutes and power reserve are indicated via vertical disks assembled coaxially. Shown in 18K white gold, Quenttin also available in 18K yellow gold and magnesium.

## JEAN-MAIRET & GILLMAN
## CESAR AUGUSTE TOURBILLON

This Cesar Auguste Tourbillon is equipped with a one-minute tourbillon with flat balance spring and offers 110 hours of power reserve. The movement beats at 21,600 vibrations per hour and features a day and night indicator, as well as day and weeks indicators. It is crafted in 18-karat gold with a sapphire crystal and caseback.

## MAURICE LACROIX

## MASTERPIECE TOURBILLON RÉTROGRADE - REF. MP7088-PL201-110_V2

The Masterpiece Tourbillon Rétrograde encased in precious platinum is limited strictly to 30 pieces and its classic tourbillon is only the first aspect of a complex trio. This combination of tourbillon and retrograde date indication alone makes this mechanical Masterpiece a sought-after rarity. The third complication is a watch movement power-reserve indication. All functions were taken into account and integrated from the start during the development of the ML 110 caliber. Engraved with the owner's name and the limited-edition issue number, the Tourbillon Rétrograde becomes a personal piece of jewelry.

## MB&F

## HOROLOGICAL MACHINE NO. 1 - REF .10.T41RL

This model displaying the hours, minutes and 175-hour power reserve (7 days) is driven by an automatic central tourbillon powered through four barrels. The 18K red-gold case (also available in 18K white gold) frames a silvered and ruthenium dial with SuperLumiNova numerals. The watch is teamed with a black hand-stitched crocodile leather strap with 18K rose-gold folding clasp, and the box contains an additional brown hand-stitched strap with 18K red-gold tang buckle.

## PARMIGIANI FLEURIER

## KALPA XL TOURBILLON CHIAROSCURO 30TH ANNIVERSARY MODEL

In 2006, Michel Parmigiani celebrated his 30th year in watchmaking with special creations like this Tourbillon 30 Seconds Chiaroscuro dedicated to Haute Horlogerie. Its manual-winding PF Caliber 501 with 30-second tourbillon, 7-day power reserve 9 (at 12:00) and 29 jewels is handmade entirely in-house and enhanced by a PVD-coloring technique; the pierced bottom plate and bridges form a mosaic of metallic colors, the result of complex laboratory research. The sapphire dial is crafted with the unique layer-by-layer electro-deposition technique. The 30-piece 950 platinum series is fitted with Hermès Havana alligator-leather straps.

## PARMIGIANI FLEURIER

## KALPA XL TOURBILLON - REF. PF011255.01

Shown in 950 platinum (and also available in rose gold), Kalpa XL Tourbillon's mechanical manual-wound Caliber PF 500.02 is crafted entirely in-house and bears 7-day power reserve (at 12:00), 28 jewels, 30-second tourbillon regulator, double barrel, hand-beveled bridges, and Côtes de Genève decoration. On a Hermès alligator leather strap and water resistant to 30 meters, this Kalpa's case is fitted with a curved, antireflective sapphire crystal and a sapphire crystal caseback, revealing the movement. Its blue/silver dial is adorned with applied indexes and delta-shaped hands. Limited to 25 pieces.

## PARMIGIANI FLEURIER

## KALPA XL TOURBILLON DIAMONDS - REF. PF011699-01

This 18K pink-gold version of the Kalpa XL Tourbilon is set with 4.943 carats of brilliant-cut diamonds. The mechanical hand-wound PF Caliber 500 with 7-day power reserve (at 12:00) and 30-second tourbillon regulator, and 29 jewels is crafted entirely in-house with hand-beveled bridges and Côtes de Genève decoration. The white mother-of-pearl dial is accented with applied 18K gold numerals and trapeze- and brilliant-cut diamonds. This Kalpa XL's Hermès calfskin strap has a mirror finish and brilliant-cut diamonds on its 18K pin buckle.

## PIAGET

## ALTIPLANO TOURBILLON GOUSSET

Piaget now revives the tourbillon pocket-watch, interpreted in a highly contemporary square form characteristic of the brand, that of the Altiplano collection. The 40mm white-gold case houses the Piaget mechanical hand-wound Calibre 600P, the world's thinnest tourbillon movement at just 3.5mm thick. Admirably reflecting the mastery inherent to the manufacture, the carriage of this tourbillon—visible through a dial opening at 9:00—is made up of 42 parts and weighs a mere 0.2 grams. Like all vintage Piaget models, the caseback is engraved with the brand's coat of arms. To provide a variety of manners of wearing it, this refined and elegant watch comes with a white-gold chain as well as a leather cordlet.

## PIAGET

### EMPERADOR TOURBILLON HAUTE JOAILLERIE - REF. GOA30018

Equipped with the mechanical manual-wind Piaget 600P shaped caliber, this high-jeweled Emperador Tourbillon is equipped with 24 jewels and 40 hours of power reserve. The three-piece case is crafted in 18-karat white gold and bedecked with 153 brilliant- and baguette-cut diamonds weighing 3.3 carats. Its dial is also ensconced in diamonds. The three bridges of the tourbillon are crafted in titanium and the movement is decorated with the Côtes de Genève pattern.

## PIAGET

## POLO TOURBILLON RELATIF - REF. G0A31123

Suspended from one end of the minute hand, the flying tourbillon carriage appears to be disconnected from the base movement driving it. The hours are read off on a central disk, while the tourbillon carriage is swept along in the rotation of the minute hand. Expressing the magic of perfect equilibrium, the Piaget Polo Tourbillon Relatif marks off time to the rhythm of the new proprietary Calibre 608P, a mechanical hand-wound flying tourbillon movement. Composed of 42 parts, the carriage—which spins on its axis once per minute, in addition to the hourly rotation around the dial—weighs a mere 0.2 grams! The new tourbillon movement, Piaget Calibre 608P, beats at a frequency of 21,600 vibrations per hour (3 Hz) and is endowed with a 70-hour power reserve.

## PIAGET

## POLO TOURBILLON - REF. GOA30111

A wonderful rendition of an iconic timepiece, this Polo Tourbillon is equipped with the 600P shaped caliber. The mechanical manual-wind movement consists of 24 jewels and offers 40 hours of power reserve. The bridges of the tourbillon are created in titanium. The case is crafted in 18-karat white gold and set with fancy-cut diamonds.

## PIAGET

## SET SKELETON TOURBILLON

The Emperador Set Skeleton Tourbillon symbolizes the alchemist's blend that Piaget achieves once again between haute horology and haute joaillerie. Piaget has developed the world's thinnest tourbillon, at a mere 3.5mm thick, and has chosen a flying version placed at 12:00 for aesthetic reasons. 159 round diamonds totaling approximately 0.2 carats and 7 sapphire cabochons (0.2 carats) adorn this exceptional caliber, whose lower side is graced with the extreme refinement of a sunray guilloché decorative motif comprising 60 divisions, which correspond to 60 seconds. This hand-wound movement has a power reserve of around 40 hours and beats at a cadence of 21,600 vibrations per hour.

## RICHARD MILLE

## 002-V2

The RM 002-V2 Tourbillon–an evolution of the first Richard Mille timepiece–houses the hand-winding caliber 002-V2 built on a carbon nanofiber baseplate. It provides hours, minutes, indicators for power reserve and torque as well as a function indicator for Winding, Neutral and Hand-setting operations. The PVD-treated variable-inertia balance has an overcoil balance spring and ceramic endstone for the tourbillon cage to provide optimal chronometric results and long-term wear. Water resistant to 50 meters, the signature tripartite Richard Mille case with sapphire caseback is offered in titanium, 18K red or white gold and platinum.

## RICHARD MILLE

## RM 014 PERINI NAVI CUP

The RM 014 Perini Navi Cup Tourbillon uses detailing inspired by nautical themes and the world of sailing. Many parts, such as the case design, crown and even the trademark Richard Mille titanium screws have undergone a transformation in this new RM model. It uses the manual-winding marine caliber 014, built on a carbon nanofiber baseplate and providing hours, minutes, indicators for power reserve and torque as well as a function indicator for Carica (winding), Neutrale (neutral) and Lancette (hand-setting). Water resistant to 50 meters, the signature tripartite Richard Mille case with sapphire caseback is offered in titanium, 18K red or white gold and platinum.

## ULYSSE NARDIN

## ROYAL BLUE TOURBILLON - REF. 799-80

A superb work, the Royal Blue Tourbillon, Ref. 799-80, is crafted in platinum and houses the UN-79 with flying tourbillon. The watch is equipped with a unique, circular rack-winding mechanism and the bridges and main plates are crafted of sapphire. The watch is available with or without a diamond bezel and with or without a diamond case. It is offered in a limited edition of 99 pieces.

## VACHERON CONSTANTIN

## MALTE TONNEAU TOURBILLON - REF. 30066

The sophisticated Malte Tonneau Tourbillon's 18K pink-gold case with curved sapphire crystal houses the mechanical manual-winding Caliber 1780 with one-minute tourbillon escapement, which has ruthenium-finished bridges. The 27-jeweled movement beats at 18,000 vibrations per hour and offers a minimum 55-hour power reserve (shown at 11:00).

## VACHERON CONSTANTIN
## OPENWORKED MALTE TOURBILLON

Housed in a stunning 18-karat pink-gold tonneau case, this Malte Tourbillon is powered by the hand-wound 1790 mechanical caliber with tourbillon regulator. It offers a 31-day analog date calendar, 55 hours of power reserve. and is completely openworked and beveled. The dial features baton hands and a crown signed with the embossed symbol of Vacheron Constantin.

## ZENITH

## DEFY XTREME TOURBILLON EL PRIMERO

Incorporating the date into the tourbillon carriage near 12:00, this Defy Xtreme Tourbillon El Primero one-minute tourbillon houses the 4035 SX automatic chronograph with 332 components. The movement's train-wheel bridge, chronograph and lower cage bridges are crafted in shock-absorbing Zenithium Z+. The watch is built with an Incabloc® anti-shock device and beats at 36,000 vibrations per hour; with automatic winding in both directions, it offers 50 hours of power reserve and times to 1/10 of a second. The heavy metal oscillating weight has carbon-fiber inserts, and the case, pushers, crown and screw-in caseback are black titanium. A helium escape valve ensures water resistance to 1,000 meters.

## ZENITH

## GRANDE CHRONOMASTER XXT TOURBILLON BLACK TIE

Cased in platinum with 60 baguette Top Wesselton diamonds weighing approximately 5.4 carats, this Grande ChronoMaster XXT Tourbillon Black Tie houses the El Primero 4005 automatic chronograph movement with one-minute tourbillon. The 319-part caliber beats at 36,000 vibrations per hour and offers 50 hours of power reserve. The chronograph measures short time intervals to 1/10 of a second. The solid gold dial has a genuine Polynesian mother-of-pearl overlay and is fitted on a handmade lambskin "Black Tie" leather strap with satin trim. This watch is water resistant to 30 meters.

## ZENITH

## GRANDE CHRONOMASTER XXT TOURBILLON EL PRIMERO CONCEPT

This limited-edition, 45mm 18K white-gold Grande ChronoMaster XXT Tourbillon El Primero Concept houses the new El Primero 4005 C automatic chronograph movement with one-minute tourbillon (revealed at 11:00). The new main plate is uniquely worked with a special design. The watch movement consists of 319 components, including 35 jewels, and beats at 36,000 vibrations per hour. The power reserve is more than 50 hours. The chronograph measures to 1/10 of a second, and offers 30-minute and 12-hour counters.

## ZENITH

## GRANDE CHRONOMASTER XXT TOURBILLON EL PRIMERO

This Grande ChronoMaster XXT Tourbillon El Primero houses the El Primero 4005 automatic chronograph movement with one-minute tourbillon at 11:00. The 319-part movement with 35 jewels beats at 36,000 vibrations per hour and offers 50 hours of power reserve. The chronograph measures short time intervals to 1/10 of a second. With bidirectional automatic winding, the watch features a central rotor on ball bearings and a 22K gold guilloché oscillating weight. The date indication is positioned around the tourbillon carriage. The 30-minute counter is at 3:00 and the 12-hour counter is at 6:00. With sapphire crystal and caseback, this version is shown in 18K rose gold and available in white gold. It is water resistant to 30 meters.

## ZENITH

## GRANDE CLASS TOURBILLON

Housing the El Primero 4035 automatic chronograph movement with one-minute tourbillon, this Grande Class Tourbillon is comprised of 325 components. Beating at 36,000 vibrations per hour, it measures short time intervals to 1/10 of a second. The Grande Class Tourbillon offers hours, minutes, tourbillon, center seconds, 30-minute and 12-hour counters. It is crafted in 18K gold and water resistant up to 50 meters.

## ZENITH

## GRANDE PORT-ROYAL TOURBILLON EL PRIMERO

Beating at 36,000 vibrations per hour, the 35-jeweled El Primero 4007 automatic chronograph movement powering this Grande Port-Royal Tourbillon consists of 298 components and offers 50 hours of power reserve. On an 18K white-gold dial, the chronograph features center seconds hand, 30-minute counter at 3:00, and 12-hour counter at 6:00. The case is black titanium with a curved, antireflective sapphire crystal and sapphire caseback. Water resistant to 50 meters, the watch is presented on a carbon-fiber integrated strap lined with calfskin.

## ZENITH

## STAR TOURBILLON EL PRIMERO

This Star Tourbillon houses the El Primero 4029 automatic chronograph movement. The 292-part tourbillon movement consists of 35 jewels, features a central rotor on ball bearings and a 22K white-gold oscillating weight with 0.3 carat of diamonds. El Primero 4029 beats at 36,000 vibrations per hour and offers 50 hours of power reserve. The 30-minute and 12-hour counters are each symbolized by a Zenith star. The 18K white-gold case, bezel and horns are set with 112 baguette-cut diamonds weighing 8.8 carats. The dial is 18K gold and mother of pearl, set with diamonds, and positioned on a black Ottoman strap.

# Equations of Time

Watches displaying the equation of time, or the difference between "true" solar time and "mean" solar time, are extremely rare. When it does appear, this function is often associated with other indications (astronomical or perpetual calendar displays, etc.). Nonetheless, the recent emergence of the first wristwatches featuring a "running" equation of time signals the return of this exclusive complication that also radiates a definite poetic

**A few astronomical notions**

There are in fact two times: true solar time and mean solar time. True solar time corresponds to the rhythms of nature. In revolving around the sun, the Earth completes an elliptical trajectory; moreover, its axis is slightly inclined in relation to the plane of the equator. The result is that the lapse of time between two successive passages of the sun past its highest point in the sky (noon) varies throughout the year. It lasts exactly 12 hours just four times per year: April 15th, June 14th, September 1st and December 24th. Otherwise, it is sometimes longer and sometimes shorter, according to a curve that is reproduced in a strictly identical manner. This difference, ranging from minus 16 minutes and 23 seconds on November 4th to plus 14 minutes and 22 seconds on February 11th, is called the "equation of time." As for mean solar time, as shown on our watches, it is in fact a pure convention based on the average of all days in the year.

1

2

### An horological challenge

The world's greatest watchmakers have vied with each other in developing ingenious systems capable of reproducing the variations in the equation of time. Since the variable lengths of true solar days are repeated in identical fashion on the same days, they may be "programmed" by means of a cam performing one complete revolution per year. The equation of time display may take on several different forms. Most watches of this kind feature a small subdial or auxiliary segment with a hand sweeping over a scale graduated from -16 to +14 minutes; the user must mentally add or subtract this difference in relation to mean time in order to know the true time. Such is the case in particular for the Equation of Time by Breguet, complete with perpetual calendar. It is also the system adopted for the Equation du temps watch by Jaquet Droz, with its sun-tipped hand running over the arc of a circle and its date and month counters.

### Running equation of time models

Distinctly more user-friendly and yet more complex in terms of construction, the so-called "running" equation–time watches have two minute hands, one indicating mean time and the other true time. This system, formerly the exclusive preserve of pocket-watches, made its first appearance on wristwatches in 2004. The Blancpain Equation Marchante is distinguished by its double equation of time display featuring a central hand adorned with a gold sun, complemented by a subdial with retrograde pointer at 2:00. The mechanism is combined with an ultra-thin perpetual calendar. Meanwhile, the Gyrotourbillon 1 from Jaeger-LeCoultre, which created a buzz mostly because of its spherical tourbillon (see the chapter on Tourbillons), is also endowed with a running equation of time featuring an additional minute hand adorned with a tiny golden sun.

3

▸▸ 1 *Breguet, Equation du temps*
*The Breguet Equation of Time displays the time between mean time and true solar time on an auxiliary segment between 1:00 and 2:00.*

▸▸ 2-5 *Blancpain, Equation marchante*
*User-friendliness is key in Blancpain's Equation Marchante, or running equation-of-time watch, and it features two minute-hands: one indicates mean solar time and the other true solar time.*

▸▸ 3 *Jaquet Droz, Equation du temps*
*Equation du temps Jaquet Droz has a sun-tipped hand indicating the number of minutes between true solar time and mean solar time, as well as date and month counters.*

▸▸ 4 *Jaeger-LeCoultre, Gyrotourbillon 1*
*The Gyrotourbillon 1 by Jaeger-LeCoultre, with its running equation of time displayed by a sun-tipped hand, created a stir with the addition of a spherical tourbillon.*

4

5

1

## Accounting for longitude

In 2005, Audemars Piguet went a step further by offering the first equation of time read-off system adjusted to the exact degree of longitude—not the entire time zone. The dial's dedicated ring enables one to determine the exact time of solar culmination (true noon) for the city engraved on the bezel. This Jules Audemars Equation of Time watch is also the first wristwatch to display sunrise and sunset times while considering three parameters: date, longitude and latitude. In that same year, the Arnold & Son brand, the legacy of British horologer John Arnold, drew inspiration from the founder's works to create an equation-time watch also accounting for longitude thanks to a dial with two mobile dial rings. This True North Perpetual watch features another original characteristic: when the 24-hour solar hand indicates true solar noon, it need only be pointed toward the sun in order for the external dial ring to indicate true geographical North.

2

▸▸ 1-2 *Audemars Piguet, Jules Audemars Equation of Time*
*The Jules Audemars Equation of Time is endowed with the first equation-of-time display adjusted to the exact degree of longitude.*

▸▸ 3 *Arnold & Son, True North Perpetual*
*The True North Perpetual by Arnold & Son boasts an equation-of-time display based on longitude and enables one to locate true geographical North.*

3

True solar time
Sun slow
Sun fast
Power Reserve
ARNOLD & SON
LONDON 1764
Time Equation
Longitude
True North

## AUDEMARS PIGUET

## JULES AUDEMARS EQUATION OF TIME - REF. 26003BA.00.D088CR.01

This Jules Audemars Equation of Time watch is created in 18K yellow gold and houses the self-winding Caliber 2120/2808. Visible through the sapphire crystal caseback, the 41-jeweled movement depicts conventional time and real solar time and offers hours and minutes, sunrise and sunset times, equation of time, perpetual calendar and astronomical moon. The solar culmination time is calibrated to the reference city chosen by the owner.

## AUDEMARS PIGUET

## JULES AUDEMARS EQUATION OF TIME - REF. 26003BC.OO.D002CR.01

This Jules Audemars Equation of Time watch is created in 18K white gold and houses the self-winding Caliber 2120/2808. Visible through the sapphire crystal caseback, the 41-jeweled movement depicts conventional time and real solar time and offers hours and minutes, sunrise and sunset times, equation of time, perpetual calendar and astronomical moon. The solar culmination time is calibrated to the reference city chosen by the owner.

## AUDEMARS PIGUET

## JULES AUDEMARS EQUATION OF TIME - REF. 25963PT.00.D002CR.0100

This Jules Audemars Equation of Time watch is created in 18K white gold and houses the self-winding Caliber 2120/2808. Visible through the sapphire crystal caseback, the 41-jeweled movement depicts conventional time and real solar time and offers hours and minutes, sunrise and sunset times, equation of time, perpetual calendar and astronomical moon. The solar culmination time is calibrated to the reference city chosen by the owner. The dial reveals the elaborate hand-engraved work.

## BLANCPAIN

## VILLERET EQUATION MARCHANTE PURE

The Equation Marchante Pure, based on the Le Brassus model that was the first wristwatch capable of displaying the separate solar and civil times by two minute-hands, was launched in 2005 as part of the Villeret collection characterized by its pure lines and formal beauty. This limited edition of 50 pieces is crafted in platinum and houses the 364-part caliber 3863A, equipped with 72 hours of power reserve.

## BLANCPAIN

## EQUATION MARCHANTE

This Le Brassus Equation Marchante is the first wristwatch with running equation of time combined with an ultra-thin perpetual calendar and a retrograde moonphase display. The dial of the Le Brassus watch is fitted with two coaxial minute hands one that indicates mean solar time and another that indicates real solar time. The watch houses Blancpain's caliber 3863, which required several years in the development process and is comprised of nearly 400 parts, including 39 rubies. Equipped with 72 hours of power reserve, the watch is presented in a 42mm platinum case. The Equation Marchante is water resistant to 50 meters and created in a limited edition of 50 pieces.

## GIRARD-PERREGAUX

## VINTAGE 1945 EQUATION OF TIME - REF. 90275

Powered by the mechanical automatic-winding GP 031Q0 caliber, this Vintage 1945 Equation of Time offers 45 hours of power reserve and displays the difference between civil and solar time, and the number of days of the current month. Additionally, the watch offers a full perpetual calendar with date, day, month, four-year cycle, and moonphase. The 30-jeweled movement beats at 28,800 vibrations per hour within a three-piece, anatomically curved 18K pink-gold case with an antireflective curved sapphire crystal. The movement can be seen through a sapphire caseback. The watch is water resistant to 3atm.

# Perpetual Calendars

Perpetual calendar watches display the date, the day, the month and generally the year, and automatically take into account the leap-year cycle. This sophisticated function requires prodigious feats of miniaturization, since the movement must have a mechanical "memory" of 1,461 days, equivalent to four years. Perpetual calendars are interpreted by all the major watch brands in original combinations of innovative mechanisms and creative display designs.

## Eternity minus one day

Astronomical watches providing the time, date, day and month—and even moonphases and signs of the zodiac—appeared in pocket watches as early as the 17th century. However, like most contemporary calendar wristwatches, these were generally "simple" calendars requiring manual correction after each month of less than 31 days, or five times a year. It was not until the late 18th century that so-called perpetual calendars were invented. The perpetual calendars took into account the full leap-year cycle: 48 months of varying lengths, including February 29th. However, while the term "perpetual" suggests everlasting accuracy, perpetual calendars actually do require a few adjustments along the route from here and eternity. They will all need to be adjusted on March 1, 2100, since the Gregorian calendar stipulates that three out of four century years are not leap years. Only a few, rare "secular calendars" that exist today will be able to automatically adjust themselves on March 1, 2100, thus saving the wearer the trouble (see the Grand Complications chapter).

2

1

3

### Different faces

Perpetual calendar models are characterized by their multiple displays (date, month, day, year, leap years), generally complemented by an indication of the lunar cycle (see Moonphases chapter). These displays may be of the pointer type or disk-based (with an aperture system). Representing a fine example of classicism revisited, the Master Eight Days Perpetual by Jaeger-LeCoultre is also the only hand-wound perpetual calendar equipped with an eight-day power reserve. A pusher on the side of the case serves to correct all the indicators in one-day increments. Offering a visual play on geometry and alongside its more traditional perpetual calendar models, Breguet's men's Classique lines up the indications of day, month, leap year and date along a vertical axis.

### Technical innovations

In its Perpetual Calendar with Under-lug Correctors, Blancpain associates classical design with technical innovation. Instead of placing the correctors along the side of the case, the brand has developed an original system enabling one to integrate them within the four wristband lugs, on the wrist side. A press of the finger is enough to adjust the date, the day, the month, the leap-year cycle or the moonphase display. The manufacture in Le Brassus also distinguished itself in 2005 by presenting the Villeret Ladies' Ultra-Slim Perpetual Calendar equipped with the "world's thinnest" ladies' perpetual calendar model (a mere 2.91mm thick). As for the Instantaneous Perpetual Calendar by Daniel Roth, it is endowed with a world-first function that instantly moves the subdials' hands at midnight. Its finely crafted self-winding movement is partially visible through the openworked dial.

### Retrograde systems

The perpetual calendar is currently providing scope for all

▸▸ 1 *Jaeger-LeCoultre, Master Eight Days Perpetual*
*The Master Eight Days Perpetual by Jaeger-LeCoultre.*

▸▸ 2 *Breguet, Classique men's watch*
*The Classique men's watch aligns the day, month, leap-year and date indications along the same vertical axis.*

▸▸ 3 *Blancpain, Perpetual Calendar with Under-lug Correctors*
*Blancpain's Perpetual Calendar with Under-lug Correctors is shown here in its 2006 Perpetual GMT version.*

▸▸ 4 *Blancpain, Villeret Perpetual Calendar Ultra-slim Ladies' watch*
*The Villeret Perpetual Calendar Ultra-slim Ladies' watch houses the world's smallest women's perpetual calendar movement (just 2.91 thick).*

▸▸ 5 *Daniel Roth, Instantaneous Perpetual Calendar*
*In the Instantaneous Perpetual Calendar by Daniel Roth, all of the calendar indications change daily on the precise stroke of midnight.*

1

▸▸ 1 *Parmigiani Fleurier, Toric Corrector Minute Repeater and Perpetual Calendar*
*The Toric Corrector Minute Repeater and Perpetual Calendar by Parmigiani Fleurier is also equipped with a retrograde date display.*

▸▸ 2 *Jaquet Droz, Perpetual Calendar*
*The Jaquet Droz Quantième Perpetual features a double retrograde display with the days on the left and the date on the right.*

▸▸ 3 *Vacheron Constantin, Openface Malte Perpetual Calendar with Retrograde Date*
*On this platinum Openface Malte Perpetual Calendar with Retrograde Date by Vacheron Constantin, the double sapphire crystal offers a spectacular view of the very heart of the mechanism.*

▸▸ 4 *Patek Philippe, Grande Complication Ref. 5104*
*To create this Grande Complication Ref. 5104 with transparent dial, Patek Philippe had to develop an original calendar-display system.*

▸▸ 5 *Patek Philippe, Annual Calendar Ref. 4937 (Patek Philippe Haute Joaillerie Ladies' watch, Ref. 4937)*
*In 2006, Patek Philippe's patented Annual Calendar was interpreted in a new Haute Joaillerie ladies' model combining white gold, mother-of-pearl and diamonds.*

▸▸ 6 *A. Lange & Söhne, Perpetual Calendar Datograph perpétuel*
*The Perpetual Datograph by A. Lange & Söhne is distinguished by its large date display, a signature feature of watches from the German brand.*

2

kinds of imaginative fancies in terms of display modes and design. So as to vary the aesthetic appeal of the dials, many watchmakers choose to use "retrograde" systems (see the chapter on Retrogrades and Jump Hours). Such is the case with the Toric Corrector Minute Repeater and Perpetual Calendar by Parmigiani Fleurier, driven by a proprietary movement and equipped with a pusher enabling instant correction of all the calendar and moonphase functions. Jaquet Droz, which is known for the original geometry of its creations, offers a Perpetual Calendar model with a double retrograde display occupying the entire vertical axis of the dial, with days on the left and the date on the right.

**A study in transparency**

Another noteworthy tendency is to play with transparency effects. Vacheron Constantin has equipped its Malte Perpetual Calendar open-face platinum watch with a double sapphire crystal front and back, affording a bird's eye view of the mechanism. The retrograde date is displayed against a highly original checkerboard-style motif on the upper part of the dial. Small subdials painted in black on the uppermost sapphire crystal indicate the day, month and leap-year cycle. Patek Philippe also played magician in 2006 by presenting its Grande

3

4

5

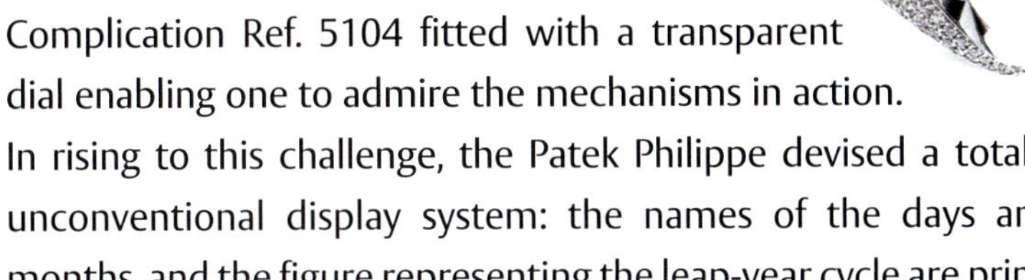

Complication Ref. 5104 fitted with a transparent dial enabling one to admire the mechanisms in action. In rising to this challenge, the Patek Philippe devised a totally unconventional display system: the names of the days and months, and the figure representing the leap-year cycle are printed directly on the sapphire crystal dial. Three tiny sapphire disks featuring black rectangular zones are placed beneath these three circles; as they rotate, these dark pallets highlight the appropriate white transfer by contrast. The platinum Grande Complication Ref. 5104 is also equipped with a minute repeater mechanism.

## Annual calendars

In 1996, Patek Philippe launched the innovative Annual Calendar, the first wristwatch capable of automatically displaying the exact month, day and date for months with 30 and 31 days and requiring just one correction per year on March 1st. All of the calendar displays are adjusted in a flash by means of the correctors recessed into the caseband. In addition to the men's models with their finely crafted dials, Patek Philippe has interpreted its patented mechanism in an exclusively feminine version lit up by shimmering mother-of-pearl and sparking diamonds. In 2006, this model was presented in an Haute Joaillerie version.

## Large date displays

Large date displays are riding a wave of unprecedented popularity that they owe both to their readability and to the touch of powerful aesthetics they lend to a dial. The German brand A. Lange & Söhne has even made the large date display one of its signature features. Displays such as those found on A. Lange & Söhne's Datograph Perpetual are based on a double-disk system with one disk for the tens and another for the second digits.

6

## A. LANGE & SÖHNE

## LANGEMATIK PERPETUAL

This self-winding A. Lange & Söhne Caliber L922.1 features oversized date, day, month, and offers four-year cycle and moonphase indication. The Sax-O-Mat movement, combined with a calendar module (Lange's own production), is covered by two patents: one for the oversized date and another for its zero-reset system. The different functions can be corrected two ways: individually by actuating the relevant correctors or simultaneously via the pusher at 10:00, which advances them all in one-day increments.

## AUDEMARS PIGUET

## ROYAL OAK PERPETUAL CALENDAR - REF. 25820SP.00.0944SP.02

A combination of 950 platinum and steel, this Royal Oak Perpetual Calendar houses the self-winding Caliber 2120/2802 with 21K gold rotor. Water resistant to 20 meters, this complication calibrates the day of the week to the date, month, phase within the 4-year leap-year cycle, astronomical moon indication with no manual adjustment required until the year 2100. It is fitted with a sapphire caseback.

## AUDEMARS PIGUET

## ROYAL OAK SKELETON PERPETUAL CALENDAR - REF. 25829ST.00.0944ST.01

This complication features an openwork dial that exposes the function and integration of the perpetual calendar, which calibrates the day of the week to the date, month, phase within the 4-year leap-year cycle, astronomical moon indication with no manual adjustment required until the year 2100. Historic base movement is hollowed and engraved by hand. The case and bracelet are stainless steel; the case's assembly screws are 18K gold. Water resistant to 20 meters, the Royal Oak Skeleton Perpetual Calendar features a sapphire crystal and sapphire crystal caseback.

## BLANCPAIN

## VILLERET LADY'S ULTRA SLIM PERPETUAL CALENDAR

This limited volume nonetheless accommodates with unfailing precision no fewer than 248 components that enable caliber 5621 to give meticulous cadence to the hours and minutes, as well as the day of the week, date, month, leap years and moonphases. This caliber is housed within an 18K white-gold case with a diameter of 34mm. Water resistant to 30 meters, the case is fitted with a sapphire crystal, glare-proof on both sides, and with a glare-proof sapphire caseback.

## BOVET

## FLEURIER SELF-WINDING PERPETUAL CALENDAR WITH RETROGRADE DATE AND SECOND TIME-ZONE INDICATION

Rather than pairing the perpetual calendar with the conventional moonphase indicator, BOVET has incorporated a more practical second time-zone function with this attractive new 42mm model shown in 18K rose gold. The perpetual calendar and 24-hour indication are combined within an exclusive mechanism constructed by BOVET. In black or white and gold, the dial displays the day, month and leap-year cycle in apertures framed by the arc of the retrograde-date scale. BOVET demonstrates its graphic skills in the creation of an uncluttered dial that conveys seven indications of time with the utmost clarity. This model is also available in 18K white gold.

## BOVET

## FLEURIER SELF-WINDING PERPETUAL CALENDAR WITH RETROGRADE DATES AND A SECOND TIME-ZONE INDICATION

Rather than pairing the perpetual calendar with the conventional moonphase indicator, BOVET has incorporated a more practical second time-zone function with this attractive new 42mm model shown in 18K white gold. The perpetual calendar and 24-hour indication are combined within an exclusive mechanism constructed by BOVET. The dial, in black or white lacquered enamel, shows the day, month and leap-year cycle in apertures, framed by the arc of the retrograde-dates scale. BOVET demonstrates its graphic skills in the creation of an uncluttered dial that conveys seven indications of time with the utmost clarity. This model is also available in 18K rose gold.

## BOVET

## SELF-WINDING PERPETUAL CALENDAR WITH RETROGRADE DATE AND MOONPHASE

This 18K rose-gold BOVET Perpetual Calendar with Retrograde Date and 55-hour power reserve is a magnificent demonstration of how to display a full calendar and moonphases via letters, numerals and images. It is easy to advance the date-hand every 24 hours against a spring released by the perpetual calendar mechanism at the end of the month so that it flicks back to the beginning. The difficulty that underlies all such retrograde mechanisms is that the hand flies back through an arc of 225° with such force that it bounces up again on the scale and catches on the second or third date. BOVET has solved this bounce-back problem by slowing the return of the date-hand with a train of wheels. (The dial is available in white or black lacquered enamel or mother of pearl.)

## BVLGARI

## BVLGARI-BVLGARI - REF. BB38PLPC

The Bvlgari-Bvlgari Ref. BB38PLPC features a mechanical, automatic winding Caliber FP 5763, with 65-hour power reserve, 35 jewels and a vibration rate of 28,800 vph. This movement offers hours, minutes, perpetual calendar and bi-retrograde seconds and is finished and decorated by hand, with stippling and Côtes de Genève treatment. The 38mm, polished platinum case is fitted with an antireflective scratch-resistant sapphire crystal, snap-on back displaying the movement through a sapphire crystal, limited-edition number engraved on the case side and is water resistant to 30 meters. The pink dial with hand-applied indexes displays hours and minutes with seconds at 6:00, leap year and month at 12:00, date at 3:00, and day at 9:00. The hand-sewn brown alligator strap has a platinum triple-fold-over buckle. Limited edition of 99 pieces.

## CHOPARD

## L.U.C LUNAR ONE WITH ORBITAL MOONPHASE, REF. 16/1894

Measuring Ø 11.4mm and 20mm thick, the L.U.C Lunar One is powered by the mechanical self-winding L.U.C 96 QP movement. The 354-part movement is a COSC-certified chronometer and bears the Poinçon de Genève quality hallmark. It has a power-reserve of approximately 65 hours and beats at 28,800 vibrations per hour. The L.U.C Lunar One features a perpetual calendar, date at 12:00, day of the week, leap year, 24-hour indication, power-reserve indicator and phases of the moon, which are presented in an extremely precise and original way. The moonphases rotate around the axis of the small seconds display, hence the term "orbital." This system's accuracy measures a mere 57.2-second difference per lunar cycle (or one day every 122 years). Water resistant to 30 meters, the watch is created in two limited editions: 250 pieces in pink gold (shown); 250 pieces in 950 platinum (Ref. 16/91894).

## CHOPARD

## L.U.C LUNAR ONE WITH ORBITAL MOONPHASE - REF. 16/91894

Measuring Ø 11.4mm and 20mm thick, the L.U.C Lunar One is powered by the mechanical self-winding L.U.C 96 QP movement. The 354-part movement is a COSC-certified chronometer and bears the Poinçon de Genève quality hallmark. It has a power-reserve of approximately 65 hours and beats at 28,800 vibrations per hour. The L.U.C Lunar One features a perpetual calendar, date at 12:00, day of the week, leap year, 24-hour indication, power-reserve indicator and phases of the moon, which are presented in an extremely precise and original way. The moonphases rotate around the axis of the small seconds display, hence the term "orbital." This system's accuracy measures a mere 57.2-second difference per lunar cycle (or one day every 122 years). Water resistant to 30 meters, the watch is created in two limited editions: 250 pieces in pink gold (Ref. 16/1894); 250 pieces in 950 platinum (shown).

## DANIEL ROTH

## PERPETUAL CALENDAR MOON PHASE - REF. 118.X.60.154.CN.BD

The Perpetual Calendar Moon Phase features an automatic movement with perpetual calendar and indication of the moonphase, offering a 45-hour power reserve and presenting a hand guilloché solid gold and platinum rotor. The two-level white-gold dial, with white enamel base and black guilloché center, displays individual counters for the indication of the days of the week, months, dates and leap years. The moonphase appears on a special counter at 12:00 with a realistic black and white image of the moon in its starry sky, represented on a moving disk. This model is water resistant to 3atm.

## DANIEL ROTH

## INSTANTANEOUS PERPETUAL CALENDAR - REF. 119.X.60.896.CN.BA

The Instantaneous Perpetual Calendar features an entirely hand-decorated automatic perpetual calendar movement with central hour and minute hand and three subdials indicating the day of the week, month, date and leap years, all jumping instantaneously at midnight. This movement is housed in a white-gold double-ellipse case, enhanced by a gold openworked dial revealing the elaborate movement decoration, further unveiled through the caseback secured by four pentagonal screws. This model is presented on a hand-stitched alligator leather strap with pin buckle bearing the brand emblem and is water resistant to 3atm.

## DE BETHUNE

## PERPETUAL CALENDAR WITH REVOLVING MOON - DB17S

The unique moonphase indication of this watch is a platinum and blued-steel sphere that emerges from a second dial turning on its own axis at 12:00. This allows for exact presentation of the stars' movements. The mechanism controlling the sphere is calculated so precisely that it will take 120 years before it will shift one day. The perpetual calendar with quick setting for the date at an intermediate position in the winding crown also features a quick setting of the month. Leap years are indicated by an eclipse of a star in a smaller aperture beneath the moonphase indication. The base movement is entirely designed and manufactured in-house. The balance is in titanium and platinum. The 43mm case is offered in rose or white gold and holds the gold dial with a hand-guilloché sun pattern. The movement offers four days of power reserve. In addition to the perpetual calendar and revolving moon indication, the watch offers minute-repeater functions via a pushpiece at 10:00.

## GERALD GENTA

## ARENA PERPETUAL CALENDAR GMT - REF. AQG.Y.87.129.CM.BD

This Arena Perpetual Calendar GMT features an entirely hand-decorated automatic movement with perpetual calendar and second time zone on a 24-hour counter at 12:00, offering a 45-hour power reserve. It is housed in a Ø 45mm titanium case with fluted caseband and platinum bezel, enhanced by a blue and satin-finish metallic net-like dial. It is water resistant to 10atm.

## GERALD GENTA

## OCTO 48-MONTH PERPETUAL CALENDAR - REF. OQM.Y.60.515.CN.BD

This Octo 48-month Perpetual Calendar offers an exclusive Gérald Genta automatic movement, entirely decorated by hand, with old-gold finish and endowed with a 48-month perpetual calendar indicating the days of the week, the date, the months of the year over a 4-year cycle, and the leap years. It is housed in an octagonal case in 5N red gold or diamond-polished, white-gold and circular satin-brushed bezel with polished hour-markers, enhanced by an engraved, cloisonné and lacquered 18K gold dial, and a beaded crown with hawk's eye cabochon. This model is presented on a black hand-stitched alligator leather strap with a circular-grained folding clasp and enhanced with the brand logo and symbol. It is water resistant to 10atm.

## GUY ELLIA

## TIME SQUARE Z1 – PERPETUAL CALENDAR

Virile and technical, the Guy Ellia style makes this watch immediately recognizable. The case is a full, simple, curved square that sits perfectly on the wrist. As for the face, it has stylized Roman numerals that create an extremely graphic universe. Overall, the watch offers strength and sobriety. On this Time Square Z1, the date, day of the week and month are indicated via hands, while the phases of the moon pass in an aperture at 6:00. The automatic Frédéric Piguet Calibre 5653 also offers leap-year indicator and three days of power reserve. It exists in white, yellow, pink and black gold and can be set with diamonds.

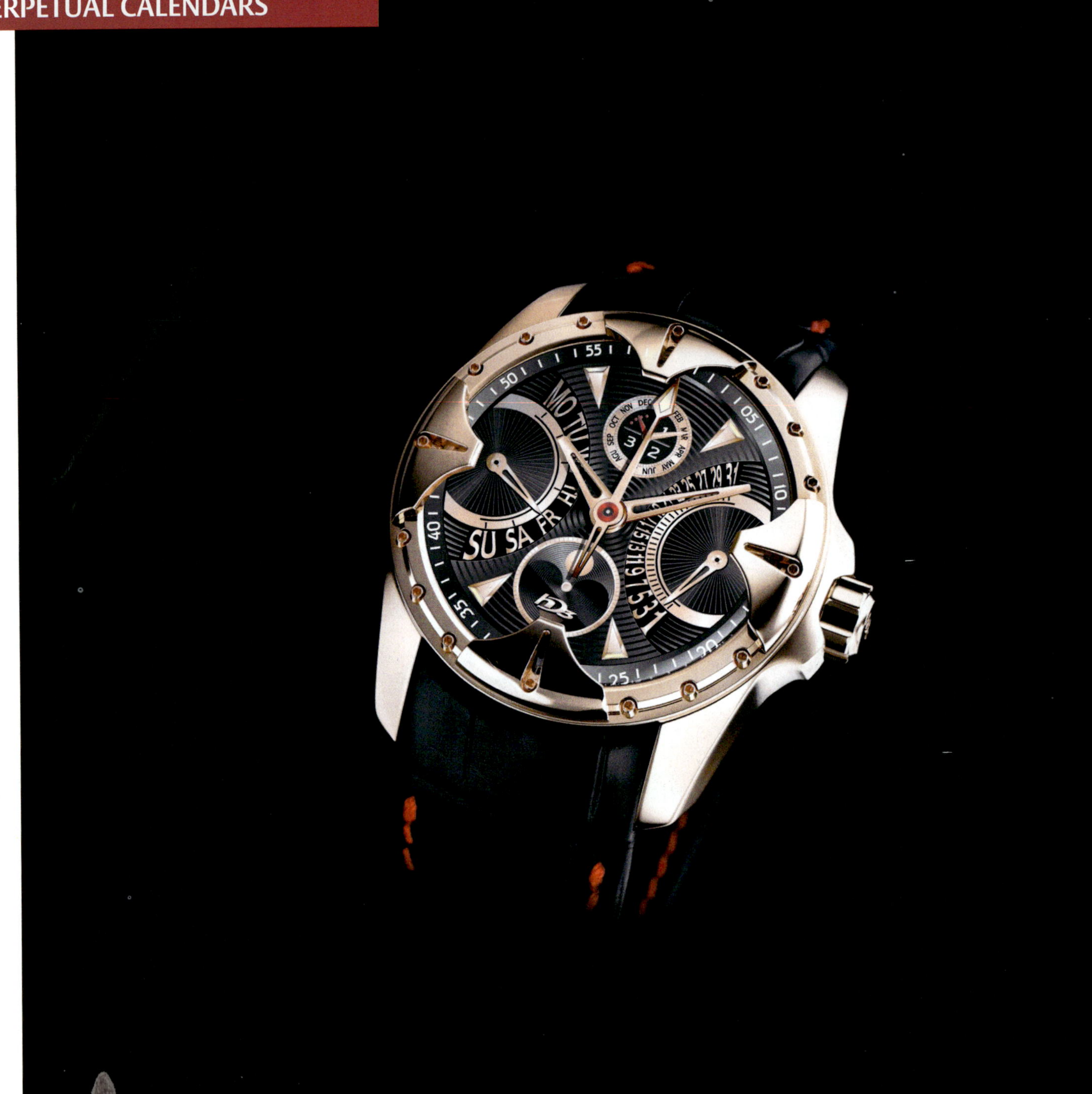

## HD3

## CAPTURE PERPETUAL

Designed by Valerie Ursenbacher, the Capture line combines technical excellence and femininity. This 43mm Capture Perpetual is powered an automatic, bi-retrograde perpetual calendar movement (visible through a sapphire crystal caseback) with 25 rubies and adjustable to 5 positions. Indications include: retrograde date at 3:00, moonphase at 6:00, retrograde day at 9:00, and month and leap year at 12:00. The dial is hand guilloché, accented with SuperLumiNova and covered with a multi-layered, antireflective sapphire crystal. Capture Perpetual is available in 6 limited editions: 4N 18K rose gold with black dial; rose gold with rose-gold dial (with or without 312 full-cut diamonds on case); 18K white gold with black dial; white gold with silver dial (with or without diamonds).

## IWC

## PORTUGUESE PERPETUAL CALENDAR - REF. IW502218

This 18K white-gold Portuguese Perpetual Calendar watch is powered by the company's 5000 caliber and offers ingenious displays. It houses the Pellaton winding mechanism with an entirely new moonphase display, a world-first patent. The Portuguese Perpetual Calendar offers seven days of power reserve and all perpetual calendar readouts. The moonphase is represented in the sky's northern and southern hemispheres, as well as via a disk display with two opposing circular windows rotating above a yellow surface with two black circles. Both moons are constantly in motion.

## JAEGER-LECOULTRE
## MASTER CALENDAR

The Master Calendar—driven by the new automatic Jaeger-LeCoultre Caliber 924—offers calendar, moonphase indicator and power-reserve indicator. The mechanical movement is crafted and decorated by hand and consists of 339 parts and 41 jewels. Beating at 28,800 vibrations per hour with 45 hours of power reserve, Caliber 924 is equipped with a 22-karat gold oscillating weight. The watch is water resistant to 50 meters and has one crown to set the hours and minutes and three correctors for the calendar.

## JAEGER-LECOULTRE

## MASTER EIGHT DAYS PERPETUAL

Classically designed, this Master Eight Days Perpetual houses the mechanical manually wound Caliber 876 that is crafted, assembled and decorated by hand. It beats at 28,800 vibrations per hour and consists of 262 parts and 37 jewels. It offers hours, minutes, date, month, day of week, four-digit year display, moonphase, power reserve, and a day/night indicator with red safety zone for changing of the perpetual calendar. The watch is crafted in platinum or in 18-karat pink gold.

## JAEGER-LECOULTRE

## MASTER GRANDE REVEIL

The Ø 43mm Master Grande Reveil offers both an alarm function (with chime or vibration modes) and perpetual calendar. Available in stainless steel, pink gold or 950 platinum, it houses the mechanical automatic Caliber 909/1 with 363 parts and and 36 jewels. offers a 24-hour indicator and moonphase indicators for both hemispheres. The 1000 Hours Control seal is engraved on the pink-gold and platinum Master Grande Reveil casebacks.

## JAQUET DROZ

## QUANTIÈME PERPÉTUEL ÉMAIL NOIR ABSOLU - REF. J008334210

The Quantième Perpétuel by Jaquet Droz is a full-fledged horological masterpiece. The interplay of hands is particularly representative of this keen sense of melding Haute Horlogerie with poetry. The serpentine hands are reserved for the Calendar functions—months, days and date—with a double-retrograde indication of the day and date, while the Lancine-type hands are dedicated to the time functions—central hour and minute indication and large seconds subdial at 6:00.

## PARMIGIANI FLEURIER

## TORIC RETROGRADE PERPETUAL CALENDAR LUNA BLU - REF. PF0011588-01

Luna Blu's perpetual calendar offers day, retrograde date, month, leap year, and high-precision moonphase in the northern and southern hemispheres. The proprietary mechanical self-winding PF Caliber 333 has 45-hour power reserve, 32 jewels, a 22K oscillating weight, hand-beveled bridges and beats at 28,800 vph. The 18K white-gold case is available with or without diamonds (1.6 carats of brilliant-cut, 3 carats of baguette-cut) and the blue lacquered 18K dial has a semi-matte circular satin-brushed rim for the date with blue transferred numerals separated by brilliant-cut sapphires. Brilliant-cut diamonds form Ursa Minor, Cetus, Lyra, and Corona Borealis. The watch comes on a Hermès crocodile leather strap.

## PARMIGIANI FLEURIER

## TORIC RETROGRADE PERPETUAL CALENDAR - REF. PF002614

This mechanical self-winding watch houses the PF333 caliber, created in house with a 22-karat gold oscillating weight and hand-beveled bridges. Beating at 28,800 vibrations per hour, the watch offers day, month, leap year, moonphase indicator and retrograde date. It is crafted in 18-karat rose gold with a slate-gray and rose-gold dial.

## PATEK PHILIPPE

## PERPETUAL CALENDAR - REF. 5140

Housed in a three-part 18K white-gold 37.2mm case, this Perpetual Calendar beats to the mechanical automatic-winding Patek Philippe 240 Q caliber. The movement, with 22K off-centered micro-rotor, features the Geneva Seal and is decorated with a Côtes de Genève pattern. The perpetual calendar offers 24-hour indication, day, date, month, year, and moonphase, plus hour and minute. Water resistant to 2.5atm, the watch features a silvered solid gold dial with applied gold markers and hands.

## PATEK PHILIPPE

## ANNUAL CALENDAR - REF. 5146/1

This superb Annual Calendar with date, day, month and moonphase takes the days of the every month except February into account. It is powered by the automatic-winding Patek Philippe 315 S IRM QA LU manufacture caliber, decorated with a Côtes de Genève pattern and beveled, bearing a 21K gold rotor and power reserve. The 18K white-gold three-piece case bears a curved sapphire crystal and the exquisite movement can be viewed through a sapphire caseback. The cream-colored dial is hand enameled and features applied gold numerals, markers and hands. The white-gold bracelet features a white-gold double-fold-over clasp.

## PATEK PHILIPPE

## ANNUAL CALENDAR - REF. 5146P

Crafted in 950 platinum, this Annual Calendar houses the proprietary mechanical automatic-winding Patek Philippe 315 S IRM QA LU manufacture caliber that is decorated with a Côtes de Genève pattern. It offers hour, minute, second, power reserve and annual calendar (date, day, month, moonphase). The calendar takes the days of a month into account over an entire year, except for February, which means it must be manually changed on March 1st. This Annual Calendar features a slate-gray dial and hand-stitched alligator strap.

## PATEK PHILIPPE

## ANNUAL CALENDAR - REF. 4937

This Calatrava cased Annual Calendar watch houses the mechanical automatic-winding Patek Philippe 315 S QA LU manufacture caliber with 21K gold rotor. Decorated with a Côtes de Genève pattern, the watch is hallmarked with the Geneva Seal. Its functions include hour, minutes, seconds, annual calendar (date, day, month, moonphase). Crafted in 18K white gold, the model is pavéd with three carats of diamonds. The dial features an engraved Calatrava Cross at 6:00.

## PATEK PHILIPPE

## CALATRAVA ANNUAL CALENDAR - REF. 5396R

Within its 18K pink-gold three-piece Calatrava case, this Annual Calendar is powered by the mechanical automatic-winding Patek Philippe 324-303 caliber. It is decorated with a Côtes de Genève pattern, features a 21K gold rotor, and is hallmarked with the Geneva Seal. The watch offers hour, minutes, seconds, day, date, month, and moonphase. It is water resistant to 2.5atm.

## PATEK PHILIPPE
## GONDOLO CALENDAR - REF. 5135P

Crafted in platinum and set with a single brilliant at 6:00, this Gondolo Calendar is powered by the automatic-winding Patek Philippe 324 S QA LU 24H caliber decorated with a Côtes de Genève pattern and beveled. The movement with its 21K gold rotor is visible through a sapphire caseback. The watch offers hour, minute and second, annual calendar and 24-hour indication. The month readout is at 2:00, moonphase with 24-hour is at 6:00, day of week is at 10:00, and date is at 12:00. The solid gold dial with slate-gray sun-patterned finish features white-gold markers and diamonds. Hallmarked with the Geneva Seal, this tonneau-shaped watch is also available in pink gold, white gold or yellow gold with different dial colors.

## PATEK PHILIPPE

## GONDOLO CALENDAR - REF. 5135R

This Gonodolo Calendar houses the mechanical automatic-winding Patek Philippe 324 S QA LU 24H manufacture caliber decorated with a Côtes de Genève pattern and featuring a 21K gold rotor. It offers hours, minute, seconds, 24 hours and annual calendar (date, day, month, moonphase, and takes the days of a month into account over a whole year, except for February). It is crafted in an 18K pink-gold two-piece tonneau-shaped case with curved sapphire crystal, pink-gold crown and screw-on caseback, displaying the movement through a sapphire crystal. It is water resistant to 2.5atm.

## PATEK PHILIPPE

## REF. 5039

The refined Clous de Paris pattern on the bezel characterizes this watch and is indicative of some models from the Calatrava collection. Its caliber 240Q with Geneva Seal is provided with a perpetual-calendar module and it is the most classic and recognizable movement among those produced by Patek Philippe. Caliber 240Q is displayed through the caseback and shows an off-center 21-karat gold micro-rotor and the exclusive Gyromax balance.

## PATEK PHILIPPE

## REF. 5059

This self-winding movement with Geneva Seal (Patek Philippe caliber 315 S-QR) offers date, day, month, four-year leap cycle and moonphase readout. The 126 module, displaying the retrograde date indication over an arc of approximately 240 degrees, is visible through the transparent caseback of the Officier case, provided with the typical hinged cover. Through the sapphire crystal caseback it is possible to observe the components of caliber 315 with its center 21-karat gold rotor engraved with the Calatrava lily-cross, a symbol of the firm and synonymous with high international prestige.

## PATEK PHILIPPE

## PERPETUAL CALENDAR

Classically executed, this Perpetual Calendar is equipped with the automatic 240 Q caliber, which beats at 21,600 vibrations per hour, is powered by an off-center, unidirectional winding 22-karat gold micro-rotor, and is visible through a sapphire crystal caseback. Due to the three calendar adjusters positioned via the lugs, the bracelet features a special mechanism on the first link—both sides—to house the extensions of the small adjustment pushers.

## ULYSSE NARDIN

## QUADRATO PERPETUAL - REF. 32090/69

The Quadrato Perpetual, Ref. 32090/69, features a 42x42mm case crafted in 18K white gold. The watch houses the UN-32 movement with perpetual calendar that is adjustable backwards and forwards from a single crown. Additionally, the watch offers second time zone on the main dial with the brand's patented quickset, and permanent home time indicated by a third hand. It is also available in 18K rose gold with various dial combinations, and is available with or without a diamond bezel.

## ULYSSE NARDIN

## GMT ± PERPETUAL - REF. 322-66/91

The GMT± Perpetual, Ref. 322-66/91, is crafted in 18K rose gold in a 42mm case. It houses the UN-32 movement that is a COSC-certified chronometer. The perpetual calendar is adjustable backwards and forwards from a single crown. The watch offers second time-zone indication on the main dial with a patented quickset. It is built in a limited edition of 500 pieces and is available in various dial combinations and on a bracelet.

## ULYSSE NARDIN

## QUADRATO PERPETUAL - REF. 326-90B/69

With the perpetual calendar adjustable backwards and forwards from a single crown, this Quadrato Perpetual, Ref. 326-90B/69, also offers second time-zone readout on the main dial with a patented quickset, and permanent home time indication via a third hand. The watch is crafted in 18K rose gold in a 42x42mm case and houses the UN-32 movement. It is available in 18K white gold with various dial combinations, and with or without a diamond bezel.

## ULYSSE NARDIN

## TELLURIUM - REF. 889-99

Crafted in platinum, the Tellurium, Ref. 889-99, houses the UN-88 movement. This astronomical masterpiece holds a host of records. The perpetual calendar completes one turn each year. The watch features a hand-painted enamel dial and is also available in 18K yellow gold.

## ULYSSE NARDIN

## LUDOVICO PERPETUAL - REF. 336-48

The Ludovico Perpetual, Ref. 336-48, is produced in 18K rose gold and houses the UN-33 movement. The perpetual calendar is adjustable backwards and forwards from a single crown. The watch offers big date in a double window. It is also available in 18K white gold with various dial combinations and is available with or without a diamond bezel.

## VACHERON CONSTANTIN

## MALTE RETROGRADE PERPETUAL CALENDAR - REF. 47031

In this Malte model, Vacheron Constantin combines a perpetual calendar with a 31-day retrograde calendar. The Ø 39mm case, crafted in either platinum or 18-karat rose gold, houses the 1126 QPR automatic movement and offers hours, minutes, perpetual calendar, day and month indications via hands. Current year and leap year are displayed through a window. The retrograde crescent offers 31-day date and the movement beats at 28,800 vibrations per hour.

## ZENITH

## GRANDE CHRONOMASTER QUANTIÉME PERPETUAL EL PRIMERO BLACK TIE

Crafted in platinum, this Grande ChronoMaster Quantième Perpetual El Primero Black Tie is powered by the proprietary El Primero 4003 automatic chronograph movement with perpetual calendar. The COSC-certified chronometer caliber consists of 321 components, including 31 jewels. With 50 hours of power reserve, the watch indicates date, moonphases, day, month and leap year, chronograph functions, and tachymeter readout. The case's 60 baguette-cut Top Wesselton diamonds (5.4 carats) encircle a solid gold dial with genuine Polynesian mother-of-pearl and 11 diamond indexes. It is released in an exclusive limited and numbered edition on a handmade lambskin "Black Tie" leather strap with satin trim.

## ZENITH

# GRANDE CHRONOMASTER QUANTIEME PERPETUAL EL PRIMERO CONCEPT

This limited-edition 18K white-gold Grande ChronoMaster Quantième Perpetual El Primero Concept houses the automatic El Primero 4003 C (a COSC-certified chronometer with perpetual calendar that beats at 36,000 vph and carries 50 hours of power reserve), which is visible through the caseback. The movement with circular-grained mainplate is composed of 321 parts including a 22K gold oscillating weight with Côtes de Genève guilloché pattern. The perpetual calendar indicates date (in ring around 30-minute counter at 3:00), moonphase (inside 12-hour counter at 6:00), day at 9:00 and month and leap year at 12:00. The sapphire dial is ringed with a tachometric scale.

## ZENITH

## GRANDE CHRONOMASTER XXT QUANTIEME PERPETUAL EL PRIMERO

This Grande ChronoMaster XXT Quantième Perpetual El Primero is a COSC-certified chronometer that houses the El Primero 4003 automatic chronograph movement with perpetual calendar. It consists of 321 components and beats at 36,000 vibrations per hour. With a 22K white-gold oscillating weight, the 31-jeweled movement offers 50 hours of power reserve and measures to 1/10 of a second. The 45mm platinum 950 PT case is water resistant to 30 meters.

## ZENITH

## GRANDE CHRONOMASTER XXT QUANTIEME PERPETUAL EL PRIMERO

This 45mm platinum 950 PT Grande ChronoMaster XXT Quantième Perpetual is framed with 60 baguette-cut Top Wesselton diamonds. Vibrating at 36,000 beats per hour, its 321-part El Primero 4003 caliber is a COSC-certified chronometer chronograph that times to within 1/10 of a second. This Grande ChronoMaster XXT offers perpetual calendar, moonphase indication, and a power reserve of 50 hours on a mother-of-pearl dial with 11 diamond indexes. The diamond-set version is available on a black leather strap accented with a strip of satin.

# Retrograde Mechanisms and Jump Hours

0
10
20
30
40
50
60
2
de GRISOGONO
GENEVE
8
SWISS MADE
15
30
60

## CEO of Urwerk

# Felix Baumgartner

### Urwerk 201

In the Baumgartner family, the passion for watches seems to be passed down through the genes. Felix Baumgartner, who founded the Urwerk brand with his brother Thomas and the designer Martin Frei, belongs to the third generation in a line of watchmakers. However, he has taken the opposite course from the tradition-based family approach, choosing instead to be open towards the present.

"This watch seems to come from somewhere else."

Upon completing his studies, Baumgartner shared the workshop of Geneva-based watchmaker Svend Andersen for a full two years. He took advantage of this period to give a certain concept time to blossom and mature, released from the constraints of a weighty horological heritage. "We are looking for ways of telling the time other than just with hands," elaborates Baumgartner. "Our basic motivation is to make watches in tune with our era and with our generation." The manufacture has therefore decided to focus on a new means of displaying time. Founded in 1995, Urwerk got off to a somewhat tentative start before presenting its 103 model at the 2003 Basel fair. Distinguished by a display mode featuring three-dimensional mobile satellites, this watch radiated a highly contemporary aesthetic appeal. A "control board" on the caseback indicated the minutes and seconds, as well as the power reserve.

Devotees of original watches were won over by the mechanical aesthetics of the 103 and immediately placed their seal of approval on the brand and followed it up with concrete orders. Since then, the annual production rate has risen from 22 watches at the time of Urwerk's launch to more than 200 in 2006. In 2005, Baumgartner cooperated with Harry Winston to create the Opus V. Based on an

*FACING PAGE*
*Felix Baumgartner, CEO, Urwerk.*

*THIS PAGE*
*The 201 in its is available in a red-gold version.*

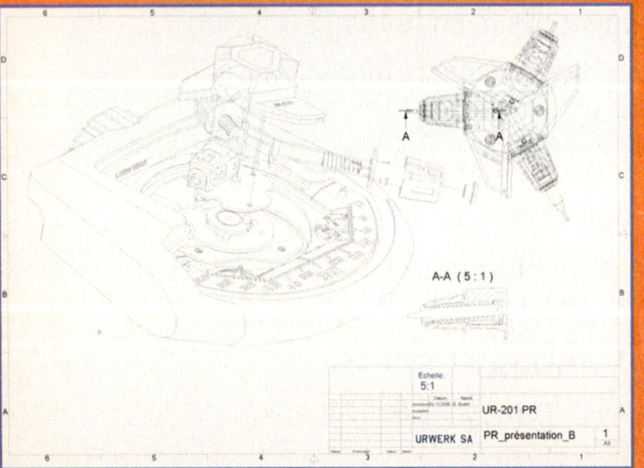


*THIS PAGE*
*Top left • The white-gold version of the 201 features raised dial numerals.*

*Center • Outline of the 201.*

*Bottom • Technical drawings of the 201.*

*FACING PAGE*
*Left • Sketches of the satellite module.*

*Right • The color of the numerals changes in the dark.*

hour display mode also featuring satellite modules, Baumgartner and Harry Winston developed a truly unprecedented watch that made a powerful impact on the 2005 watchmaking vintage. In addition to the its distinctive satellite hours, the Opus V also contained a retrograde minute indication, as well as day/night and power-reserve displays. A 0 to 5 year graduation on the caseback warns the owner when his watch is due for servicing.

An evolution of the Opus V, the Urwerk 201 was presented in early 2007. Similar and yet quite distinct, the latest arrival features the same fundamental complication—the Urwerk patented satellite hour display—but the retrograde hand of the Harry Winston model is replaced by a pointed telescopic or retractable pointed extension that withdraws inside a cylinder when it is not sweeping over the minute dial. Each branch of the satellite triangle thereby houses one of these telescopic hands. "This watch is more complicated than the 103 in terms of design and more so than the Opus V in technical terms," notes Baumgartner. The 201 is endowed with a power-reserve display in the lower left-hand part of the dial and a day/night indicator on the right. The all-black case middle, topped by a sapphire crystal, is given an extremely fine micro-sandblasted treatment before it is diamond-polished with a machine—an operation that simply cannot be done by hand. As with all Urwerk models, the caseback is titanium. Fitted with a carousel that the watchmaker himself describes as "imposing," the 201 features various fine-tuning adjustments on the back, such as the quinquennial service indication and even a display keeping track of the years for which the movement has been running—a first in watchmaking. Urwerk expects to produce a maximum of 50 such watches annually and delights in creating watches that are actually wearable. "We don't want to be part of the trend for monster timepieces. Although fairly reasonable, the dimensions of the 201 nonetheless give the distinct impression that this watch comes from somewhere else," explains Baumgartner.

# Retrograde Mechanisms and Jump Hours

Certainly not all hands must turn steadfastly around dials. Over the past few years, so-called "retrograde" and "jumping" displays have been making a noteworthy comeback. Often combined in various ways, they facilitate the creation of original and lively dials. Several such creations, some devised by independent watchmakers, show that boldness and imagination are very much the name of the game when it comes to original displays.

## Back to the future

From a relatively early stage in horlogical history, watchmakers attempted to wander off the beaten track by developing display modes different from the standard hands spinning completely around an axis. They explored ideas such as retrograde systems in which hands cover the arc of a circle and then returning instantly to their point of departure in order to start their trajectory all over again. These days, retrograde displays are making a much-publicized comeback; they enable a brand to distinguish itself by composing dials with extremely original designs. Witness the Sectora II Automatic by Jean d'Eve, which displays the hours and minutes across a 120° segment—a geometrical shape that inspired the creation of an original case design. In a more classic aesthetic vein, the Double Rétrograde II by Maurice Lacroix, in addition to "traditional" hours, minutes and seconds, features retrograde date and 24-hour displays. In 2006, Harry Winston merged Haute Horlogerie and Haute Joaillerie in its diamond-set Ocean Lady Biretro. The mother-of-pearl dial is adorned with a double retrograde display of the day and the seconds, in addition to off-centered hour/minute and date indications.

1

▸▸ 1 *Jean d'Eve, Sectora II Automatic*
*Jean d'Eve plays with geometric effects in its Sectora II Automatic.*

▸▸ 2 *Maurice Lacroix, Double Rétrograde II*
*The Double Retrograde II by Maurice Lacroix boasts a retrograde date display and a 24-hour dual time-zone indication.*

▸▸ 3 *Harry Winston, Ocean Lady Biretro*
*Harry Winston's Ocean Lady Biretro has retrograde day and seconds.*

▸▸ 4 *Pierre Kunz, Seconde Virevoltante Rétrograde*
*On the Seconde Virevoltante Rétrograde by Pierre Kunz, the seconds hand makes a 417° loop before returning to its initial position in five hundredths of a second.*

▸▸ 5 *Milus, Herios Triretrograde Seconds Skeleton*
*Herios Triretrograde Seconds Skeleton by Milus, with three seconds-hands running over three 20-second segments.*

▸▸ 6 *Gérald Genta, Fantasy Racing*
*On the Gérald Genta Fantasy Racing model, Mickey's arm serves as a retrograde minute hand.*

## A spectacular show

The dance of hands leaping back to their initial position often gives retrograde timepieces a particularly stunning appearance. Especially when watchmakers make a point of complicating these "figures." On the Seconde Virevoltante Rétrograde model, Pierre Kunz offers a fascinating "looping the loop" sequence inspired by aerobatics. The seconds hand dips and soars around a 417° orbit, before returning to its initial position in 5/100ths of a second and starting off again. The Herios TriRetrograde Seconds Skeleton by Milus features a retrograde seconds display on three 20-second segments placed at 6:00, 10:00 and 2:00: the first hand moves over one segment, then returns instantly to its starting point as the next one takes over, before handing over to the third.

## Jump hours

Retrograde systems are often associated with jump-hour displays where the hand is replaced by a disk bearing the hour numerals that "jumps" one notch forward every hour. Such is the case with the entertaining Fantasy Racing model by Gérald Genta, equipped with retrograde minutes and featuring Mickey's arm doubling as a hand. In 2006, de GRISOGONO presented an original option with its FG One. The elongated case features two dials, one with retrograde minutes over a 230° segment, instant jump hours and a "dragging" second time zone; the other dial bears retrograde seconds complete with a day/night indication synchronized with the second time zone. Meanwhile, the Digitale from De Bethune has an all-disk display mode through an upper day/date/month aperture, a middle aperture for minutes, and a lower one for jump hours—all arranged around a dial adorned with Côtes de Genève and jewels evoking the finest mechanical movements.

2
MAURICE LACROIX
Haut
Bas
SWISS MADE
3
Gérald Genta
GENEVE
4
GENEVE
PIERRE KUNZ
5
HARRY WINSTON
MON TUE WED THU FRI SAT SUN
SWISS MADE
6
MILUS

▸▸ 1 de GRISOGONO, FG One
de Grisogono combines jumping hours, retrograde minutes and a dragging dual time-zone display on the original FG One fitted with two dials.

▸▸ 2 De Bethune, Digitale
The Digitale by De Bethune is distinguished by an entirely disk-based display.

▸▸ 3 HAUTLENCE, HL06
A revolutionary jumping-hour display was designed for the HL06 by HAUTLENCE.

▸▸ 4-5 Harry Winston, Opus 3
For the Harry Winston Opus III, Vianney Halter devised an entirely numerical display with six portholes and 10 independently rotating disks.

▸▸ 6 Daniel Roth, Ellipsocurvex Papillon
The Ellipsocurvex Papillon by Daniel Roth has a clean, layered display.

1

2

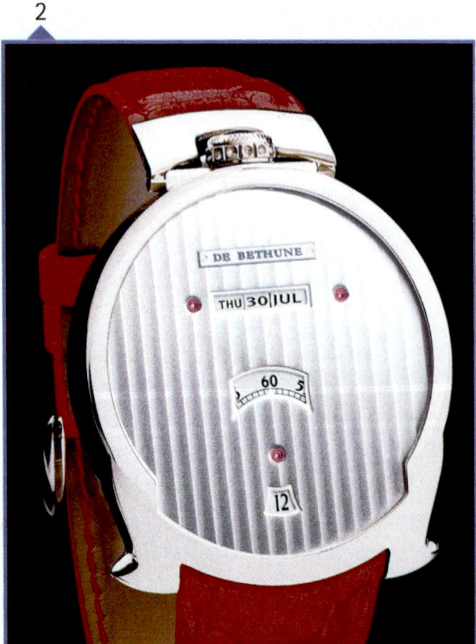

1

*Vincent Calabrese, Baladin*
*Vincent Calabrese's Baladin features an original read-off system referred to as a "wandering hour."*

## New faces of time

Numerical mechanical displays are much in favor with watchmakers at the moment, who see this as a fascinating field of exploration providing scope for all manner of fanciful inventions. The youthful Neuchâtel-based brand Hautlence offers a unique jump-hour display with a perforated upper disk moving in jumps over a numbered plaque and controlled via jointed rods that are visible through the openworked dial. With his Baladin, Vincent Calabrese offers a thoroughly unconventional read-off system he poetically calls "wandering time." The jump-hour numerals appear through a small round aperture, which itself spins around the dial to mark off the minutes, while a large central hand indicates the seconds. For the Opus III by Harry Winston, Vianney Halter devised an entire numerical display with hours, minutes, seconds and jump hours appearing in six separate "portholes." The 10 independently rotating disks are driven by two mechanisms comprising a total of 250 parts.

## The third dimension

Another innovative approach consists of three-dimensional displays. Daniel Roth set the tone with its Ellipsocurvex Papillon, a watch with two minute hands positioned at an 180° angle. Upon reaching the number 60, the first hand pivots 90° and disappears beneath the hidden part of the dial, while the other appears at 00 minutes via the same pivoting mechanism. Upon the ultra-futuristic Urwerk 103 Blackbird, hands are replaced by four cones, each displaying three different hours. The cross bearing the cones turns in a clockwise direction, while the cone spins on its axis like a satellite. By sliding along a scale graduated from 1 to 60, the hour cube also points to the minutes. This fascinating mechanism may be admired in action through a sapphire crystal. In 2006, the American brand Jacob & Co caused a sensation by introducing the Quenttin with a

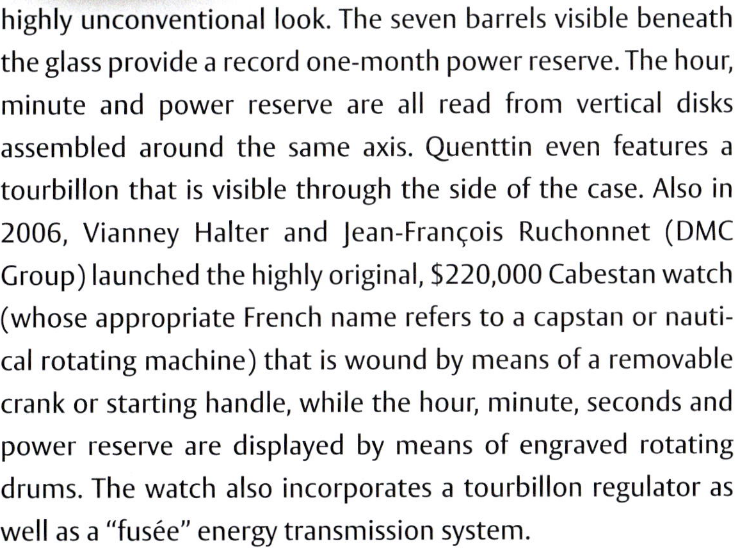

highly unconventional look. The seven barrels visible beneath the glass provide a record one-month power reserve. The hour, minute and power reserve are all read from vertical disks assembled around the same axis. Quenttin even features a tourbillon that is visible through the side of the case. Also in 2006, Vianney Halter and Jean-François Ruchonnet (DMC Group) launched the highly original, $220,000 Cabestan watch (whose appropriate French name refers to a capstan or nautical rotating machine) that is wound by means of a removable crank or starting handle, while the hour, minute, seconds and power reserve are displayed by means of engraved rotating drums. The watch also incorporates a tourbillon regulator as well as a "fusée" energy transmission system.

▸▸ 7-8 *Urwerk, 103 Blackbird*
*On the ultra-futuristic 103 Blackbird by Urwerk, four cones (each of which indicates three different hours) replace hands.*

▸▸ 9 *Jacob & Co, Quenttin*
*Also a tourbillon, Jacob & Co.'s Quenttin displays the hour, minute and 31-day power reserve on vertical disks assembled along the same axis.*

## BOVET

## FLEURIER JUMPING HOUR OPENWORK 42MM

BOVET's most unusual timepiece features a central jumping digital hour set in a rock crystal glass revealing the poetry of its original movement. The white of mother-of-pearl rotating ring shows the minutes against the cabochon sapphire set on the dial ring. The 12 central hours are all on one disk, offset so that its rim is leveled with the movement's rim. They jump instantaneously as the 60th minute reaches the dot at 12:00. A sapphire crystal caseback reveals the movement and its gold winding rotor, engraved in fleurisanne style. The jump hour watches are also available in 18K white gold; with a 16-point Compass Rose painted in different colors on a mother-of-pearl or fired-enamel dial; or in a jewelry version upon which the Compass Rose is paved in diamonds and sapphires.

## BOVET

## FLEURIER COMPASS ROSE JUMPING HOUR

Shown in 18K white gold, BOVET's 42mm Fleurier Compass Rose Jumping Hour features a central jumping digit set in a stylized Compass Rose encircled by a rotating ring showing the minutes. The 12 central hours are placed on one disk, offset so that its rim is leveled with the movement's rim. The hour jumps instantaneously as the 60th minute reaches north on the rose. The 16-point rose—eight points for the major winds and eight for the half winds—is hand-painted on the mother-of-pearl. A sapphire crystal caseback reveals the movement and its gold winding rotor engraved in fleurisanne style. The watch is available in rose gold, designed with a diamond-and-sapphire Compass Rose, or with an openwork dial.

## BOVET

## FLEURIER COMPASS ROSE JUMPING HOUR

This jewelry version of BOVET's 42mm Fleurier Compass Rose Jumping Hour is shown in 18K white gold. The central jumping hour is set in a stylized Compass Rose encircled by a rotating ring showing the minutes. The 12 central hours are all placed on one disk, offset so that its rim is leveled with the movement's rim. The hours jump instantaneously as the 60th minute reaches north on the rose. The sapphire-and-diamond 16-point rose—eight points for the major winds and eight for the half winds—makes for a striking design on the mother-of-pearl dial. A sapphire crystal caseback reveals the movement and its gold winding rotor engraved in fleurisanne style. BOVET's Jumping Hour models are available in 18K rose gold and with miniature enameling or openwork dials.

## BOVET

## 39MM ROSE-GOLD JUMPING HOURS WITH MINIATURE PAINTING

This Madonna and Child replica from Raphael's original 16th-century painting is executed on a mother-of-pearl dial and the unique piece required more than 70 hours of intense work to create. The Madonna and Child Jumping Hours was sold at auction in 2005 during the Monaco Only Watch event hosted by Antiquorum. The entire amount of the sale was given to the Monegasque Association against Duchenne Muscular Dystrophy. The watch houses the self-winding caliber 11BA07 by Vaucher Manufacture, which was developed according to Bovet's specifications. It consists of 32 rubies, twin barrels and a 22-karat gold winding rotor engraved in Fleurisanne style. It beats at 28,800 vibrations per hour and offers 55 hours of power reserve. The bridges are beveled and decorated in the Côtes de Genève pattern.

## BREGUET

## CLASSIQUE – REF. 5207BB/12/9V6

This Classique watch is crafted in 18K white or yellow gold, housing a self-winding movement with a 65-hour power reserve. Its engine-turned and silvered gold dial features off-center hours and minutes, a power-reserve indicator at 12:00 and retrograde small seconds at 6:00. The movement is protected by a water-resistant case with sapphire crystal caseback.

## BREGUET

## TRADITION - REF. 7037BA/11/9V6

This Tradition wristwatch is crafted in 18K yellow or white gold and houses a self-winding movement. Its dial is in silvered 18K gold, hand-engraved on a rose-engine, off-centered at 12:00 and displays a chapter ring with Roman numerals. Retrograde small seconds are engraved on the movement. Its case is enhanced with a sapphire caseback.

## BVLGARI

## ASSIOMA - REF. AAP48GLHR

The Assioma features an in-house manufactured, mechanical, automatic-winding Caliber BVL 261. This movement, consisting of 261 pieces offers a 45-hour power reserve, 44 jewels and a vibration rate of 28,800 vph, is finished and decorated by hand, including a stippled anthracite back and a skeletonized rotor with engraved ring. Its functions include off-centered retrograde hour (240°), minutes, AM/PM indication and small seconds. The curved, 48mm, polished 18K pink-, yellow- or white-gold case, fitted with a curved, antireflective, scratch-resistant, sapphire crystal, an 18K pink-gold crown, and a snap-on back displaying the movement through a sapphire crystal, bears the limited-edition number engraved on its side and is water resistant to 30 meters. The in-house manufactured dial has hand-applied indexes, and opaline silver central and external decorations, circular satin-finished anthracite hour and minute areas and an opaline small second circle. Its display includes AM/PM Indication and small seconds at 6:00. The pre-curved, rolled-edge, brown alligator strap has an 18K matching gold triple-fold-over buckle. Limited edition 99 pieces.

## DANIEL ROTH

## ELLIPSOCURVEX PAPILLON - REF. 318.Y.70.351.CM.BD

Available in white gold, 5N red gold or platinum, this Ellipsocurvex Papillon features a unique system with two retractable hands that are positioned at 180-degree angles to one another and that always turn in the same direction on a single axis. The minute track is a U shape that runs from 00 to 60 between 3:00 and 9:00. On reaching the 60 position, one of the hands pivots 90 degrees on its own axis and disappears behind the hidden part of the dial, while the other hand appears at the same time on the dial at 00 minutes—activated by the same pivoting mechanism. This exclusive patented movement is hand-beveled with Côtes de Genève finishing.

## GERALD GENTA

## ARENA SPORT BIRETRO - REF. BSP.Y.80.265.RV.BD

The Arena Sport Biretro offers the exclusive Gérald Genta biretro self-winding movement, entirely hand-decorated, "Potter" finishing, equipped with SuperLumiNova jumping hours through an aperture at 12:00, a 210° retrograde minutes segment in the upper part of the dial between 8:00 and 4:00, and a 180° retrograde date segment on the lower part of the dial between 4:00 and 8:00. It is housed in a Ø 45mm titanium case, enhanced by a green and white dial on a green rubber strap. It is water resistant to 10atm.

## GERALD GENTA

## ARENA BIRETRO GOLD - REF. BSP.Y.66.269.CN.BD

The Arena Biretro Gold offers the exclusive Gérald Genta biretro self-winding movement, entirely hand-decorated, "Potter" finishing, equipped with jumping hours through an aperture at 12:00, a 210° retrograde minutes segment in the upper part of the dial between 8:00 and 4:00, and a 180° degree retrograde date segment on the lower part of the dial between 4:00 and 8:00. It is housed in a polished white-gold case with fluted middle, brushed white-gold crown guard and circular satin-brushed tantalum bezel. It is enhanced by a multi-layer tantalum colored metallic dial with circular satin-brushed finish, polished applied minute numerals, and retrograde date segment with a sunray guilloché motif. This model is water resistant to 10atm.

## GERALD GENTA

## ARENA BIRETRO SET - REF. BSP.Y.86.999.CA.BD.S75

This version of the Arena Sport Biretro offers the exclusive Gérald Genta biretro self-winding movement, entirely hand-decorated, "Potter" finishing, equipped with jumping hours, retrograde minutes, and retrograde date. It is housed in a titanium case with fluted middle and white gold bezel set with 76 rubies (0.74 carat). This original model is enhanced by a jewelry dial available with 28 diamonds (0.13 carat) and 99 rubies (0.41 carat). This model is presented on a Gérald Genta black rubber strap with folding clasp and is water resistant to 10atm.

## GERALD GENTA

## OCTO BIRETRO - REF. OBR.Y.50.510.CN.BD

The Octo Biretro features the exclusive Gérald Genta biretro hand-decorated automatic movement with a jumping hour and retrograde minutes and date, offering a 45-hour power reserve. It is housed in a Ø 42.5mm or Ø 39mm (X size) octagonal 5N red-gold or white-gold case, enhanced by a black and red or ivory and black lacquered dial. This model is presented on a black alligator strap with folding clasp and is water resistant to 10atm.

## GERALD GENTA

## FANTASY AVIATOR - REF. RSF.X.10.143.LB.BA

The Fantasy Aviator presents with originality the exclusive Gérald Genta retro automatic movement with jumping hours and retrograde minutes, offering a 42-hour power reserve and housed in a Ø 41mm steel case with fluted caseband. The originality of this piece resides in its fantasy dial presenting a Mickey Mouse aviator motif with an articulated arm showing the retrograde minutes. This model is presented on a brown leather strap and is water resistant to 3atm. Fantasy Aviator is also available with black rubber bezel and strap.

## GIRARD-PERREGAUX

## CAT'S EYE BI-RETRO - REF. 80485D52A761

Powered by the mechanical automatic-winding GP 03390 caliber (GP 3000 base and moonphase and retrograde weekly-calendar module), this Cat's Eye Bi-Retro offers hours, minutes, day, date, moonphase and power reserve. The watch is crafted in 18K pink gold and the bezel is set with 48 diamonds framing a stunning mother-of-pearl dial with two retrograde indications and moonphase. The 36-jeweled GP 03390 beats at 28,800 vibrations per hour and offers 46 hours of power reserve. Its bridges are beveled and decorated with a Côtes de Genève pattern and the pillar-plate with a circular-graining pattern. The movement is visible via a sapphire caseback.

## GUY ELLIA

## TIME SQUARE Z1 – JUMPING SECOND

Masculine and technical, Guy Ellia's style makes this watch immediately recognizable. The case is a full, simple, curved square that sits perfectly on the wrist. It has stylized Roman numerals that create an extremely graphic universe offering strength and sobriety. Set in a semi circular counter at 6:00, the jumping retro-second hand comes back around to its starting point every thirty seconds (automatic mechanism by Frédéric Piguet). With pure pleasure the wearer can contemplate the passage of time, from half-minute to half-minute. This watch is created in white, yellow, pink or black gold and can be set with diamonds.

## HARRY WINSTON

## PREMIER BIRETROGRADE

Designed in the New York studios and hand crafted and assembled by a master watchmaker in Switzerland, the Premier Biretrograde watch features an automatic movement consisting of 223 parts and 36 rubies. A double-retrograde module designed exclusively for Harry Winston has been adapted to work with the Girard-Perregaux GP 3106 automatic movement. This outstanding watch features retrograde display of both seconds and the day of the week at 9:00 and 3:00. The 35mm case is crafted in 18K white or yellow gold and is water resistant to 30 meters. The watch required 14 months to make and features a sapphire caseback.

## HARRY WINSTON

## OCEAN LADY BIRETRO

The Ocean Lady Biretro offers two retrograde readouts, one for seconds and one for days of the week. Indicated via SuperLumiNova hands, the hours are off-centered at 12:00. The 36mm 18K gold case houses the mechanical self-winding GP 3106 movement with an openworked 18K rose-gold rotor. The watch is water resistant to 100 meters. Shown with black rubber straps, it is available on pink or pale blue rubber straps. The Ocean Lady Biretro is available in a gem-set version with 186 brilliant-cut diamonds on the dial and bezel, for a total of 2.76 carats.

## HAUTLENCE

## HL07

Housed in a 43.5x37mm 18K pink-gold case HL07 is powered by a Hautlence hand-winding mechanical caliber with 40-hour power reserve and beating at 21,600 vph. The silvered opaline dial with sunray design and SLN C3 luminescent numerals displays offset skeleton subdials for the jumping hours, retrograde minutes, and seconds. The hand-sewn alligator leather strap has either an 18K pink-gold adjustable triple fastener with double-safety pushbutton or pin buckle. Water resistant to 3atm, HL07 is fitted with a bevel-edged sapphire crystal and a sapphire crystal caseback. Limited to 88 pieces, each HL07 has an identification plate and is numbered.

## HAUTLENCE

## HL08

Housed in a 43.5x37mm 18K gray-gold case HL08 is powered by a proprietary hand-winding mechanical caliber with 40-hour power reserve and beating at 21,600 vph. The silvered opaline dial with sunray design and SLN C3 luminescent numerals displays offset skeleton subdials for the jumping hours, retrograde minutes, and seconds. The hand-sewn alligator leather strap has either an 18K gray-gold adjustable triple fastener with double-safety pushbutton or pin buckle. Water resistant to 3atm, HL08 is fitted with a bevel-edged sapphire crystal and a sapphire crystal caseback. Limited to 88 pieces, each HL08 has an identification plate and is numbered.

## HD3

## IDALGO - XT1

HD3's Idalgo line is created by Jorg Hysek. Exclusive to HD3 Idalgo XT-1's movement has 32 rubies, 42-hour power reserve and beats at 28,800 oscillations per hour. Idalgo XT-1 displays the jumping (or flying) hour aperture at 3:00, $180^{\circ}$ retrograde 60-minute hand (with SuperLumiNova) and $360^{\circ}$ retrograde seconds hands, and date on left. Fitted with an antireflective, tinted sapphire crystal, the ergonomic case adapts to fit different wrist sizes. Idalgo XT-1 is available in 5 33-piece limited editions: 4N 18K rose gold with black dial; titanium and rose gold with black or grained dial; titanium and 18K white gold with black dial or grained dial.

## HD3

## TRINITY

The three designers of HD3 pooled their talents to create Trinity. Its automatic movement is unique to HD3, holds 20 rubies, 40 hours of power reserve, has a frequency of 28,800 alterations per hour, and a balance lift of 50°. The oscillating rotor was created by HD3 and adds new dimension to the movement. Trinity's innovative display employs 3 rotating disks for the hours, minutes, and seconds. Several limited editions are released: 333 pieces in titanium with white or pink gold; 33 pieces in pink gold; and the black titanium-cased Black Magic series of 99 pieces: 33 with yellow, 33 with orange, and 33 with green luminova numerals.

## PIAGET

## RECTANGLE A L'ANCIENNE XL - REF. GOA29116

Created in pink gold, this oversized rectangular watch houses the Piaget 561P caliber and offers a power reserve indicator and a retrograde seconds indicator. The case is set with 172 round diamonds and the dial is set with 88 diamonds.

## URWERK

## 201 – HAMMERHEAD

The orbital karrusel is the key component of the complication. It is responsible for not only the rotation of the hour satellites but also the integral telescopic minute hands extending and retracting through the middle of the revolving hour satellites. The telescopic minute hands, when extended, enable the 201 to display the time on a large, easy-to-read, dial. They then retract to allow for a very wearable and comfortably sized case.

## VACHERON CONSTANTIN

## MALTE OPENFACE RETROGRADE PERPETUAL CALENDAR - REF. 47032

This platinum Malte Openface Retrograde Perpetual Calendar houses the mechanical automatic-winding Vacheron Constantin Caliber 1126 QPR. The 36-jeweled movement beats at 28,800 vibrations per hour, has 38 hours of power reserve, is hand-finished, and features a pink-gold mainplate. All of its parts have been hand-chiseled and etched away for exceptional viewing. The perpetual calendar offers date, day, month, digital year indication, and 4-year cycle. The dial consists of a sapphire crystal disk, luminescent peripheral date-ring, and white-gold hands. On an alligator strap with platinum clasp on an alligator strap with platinum clasp.

## VACHERON CONSTANTIN

## PATRIMONY BI-RETROGRADE - REF. 86020

Powered by the mechanical automatic-winding Vacheron Constantin Caliber 2460 R31 R7, this Patrimony Bi-Retrograde is hallmarked with the Geneva Seal. Beating at 28,800 vibrations per hour and offering 43 hours of power reserve, the 27-jeweled movement is equipped with a guilloche pink-gold rotor. Retrograde day is provided at 6:00 and retrograde date is shown at 12:00. The 18K pink-gold three-piece case is high-polished and water resistant to 3atm. The sapphire caseback reveals the exquisite movement.

## ZENITH

## CHRONOMASTER OPEN EL PRIMERO RETROGRADE

This ChronoMaster Open El Primero Retrograde houses the El Primero 4023 automatic chronograph movement with retrograde date indicator from the 30-minute counter axis. The 50-hour power-reserve indicator is at 6:00. The 39-jeweled movement consists of 299 components and beats at 36,000 vibrations per hour. The watch offers 30-minute counter at 3:00, a 3-branched small seconds hand at 9:00, central chronograph seconds hand, and measures short time intervals to 1/10 of a second. It is offered in 18K rose gold with opened Grain d'Orge guilloche-patternedguilloché-patterned dial at 10:00 to view the El Primero movement.

# Multiple time zones and GMTs

In their quest to display times simultaneously in several time zones, watchmakers have invented all kinds of systems ranging from the double hour-hand to the so-called "world time" display. Boosted by the ongoing rise of intercontinental travel, globalization of economies and the demands inherent to modern communications, this extremely useful function is attracting an unprecedented wave of interest and inspiring all manner of innovations in terms of both mechanics and display modes.

## Around the world in 24 hours

Until the late 19th century, each region had its own local time loosely based on longitude. As long-distance travel became increasingly common with the construction of railways, the need to establish a structured time system became increasingly apparent. The International Meridian Conference in 1884 delegates divided the planet into 24 time zones, using the Greenwich Meridian as the reference point (Longitude 0). The first dual-time zone watches were equipped with two fairly coordinated movements displaying the time on separate dials. This early device was perfected by using a single movement with a disconnecting-gear system for the second hour-hand. These days, watchmakers are vying with one another to find ingenious ways to develop increasingly original systems for watches with two, three, four or more time zones.

1

3

2

### Pointer-type systems

The simplest and most common system involves the addition of a second hour-hand distinguished by its shape or color, but this principle may be interpreted through countless variations. On the Malte Dual Time Regulator from Vacheron Constantin, the minutes are shown with a central hand, the main hour in a small dial at 12:00 and the second time-zone on a 24-hour subdial at 9:00. Les Longitudes by Jaquet Droz is also inspired by regulator-type dials with central minutes, one subdial with Roman numerals for the first time, and another with Arabic numerals for the second 24-hour time zone. For its L.U.C Pro One GMT, Chopard devised an unconventional read-off system: When the user wishes to view the time in another part of the world, he simply turns the bezel with its large luminescent orange numerals so as to line up the second time zone with the triangular hand. Franck Muller has spared a thought for countries such as India, which is five and a half hours ahead of Greenwich Mean Time, by displaying coordinated time zones on the dial of its Master Banker II, including one that is adjustable to the corresponding half hour. Meanwhile, Piaget's Altiplano Double Jeu reinvents the so-called "secret" watches by means of two superimposed cases linked by a hinge and driven by the mechanical ultra-thin movements that are the recognized specialty of the manufacture.

### One hand may conceal another

In most pointer-type systems, the hand of the second time zone remains permanently visible, even when not required by the user. In 2001, Patek Philippe solved this aesthetic issue in an ingenious manner. The Calatrava Travel Time has two hour-hands that may be superimposed. Simply pressing one of the pushers recessed into the case side "detaches" the local hour hand and moves it forwards or backwards in one-

4

5

▸▸ 1 *Vacheron Constantin, Malte Dual Time Regulator*
*Malte Dual Time Regulator by Vacheron Constantin has central minutes, main hour at 12:00 and 24-hour dual time-zone display at 9:00.*

▸▸ 2 *Patek Philippe, Calatrava Travel Time*
*Calatrava Travel Time by Patek Philippe is fitted with two superimposed hands and second-time-zone adjustment using + or – pushers.*

▸▸ 3 *Jaquet Droz, Les Longitudes*
*On the Jaquet Droz Les Longitudes watch, the first time zone appears in Roman numerals and the second in Arabic numerals on a 24-hour scale.*

▸▸ 4 *Piaget, Altiplano Double Jeu*
*The Altiplano Double Jeu by Piaget consists of two superimposed cases housing two ultra-thin mechanical hand-wound movements.*

▸▸ 5 *Chopard, L.U.C Pro One GMT*
*For its L.U.C Pro One GMT, Chopard devised an original read-off system based on a rotating bezel.*

hour increments, convenient for traveling eastward or westward. When the dual time-zone function is not in use, the two superimposed hands sweep round the dial in a perfectly synchronized fashion. The Master Hometime by Jaeger-LeCoultre has a fairly similar double-hand system, but which is activated exclusively by the single crown at 3:00.

### Aperture-based systems

There are also some multiple time-zone watches equipped with aperture-type systems, which also gave rise to a wide range of technical and aesthetic variations. Jaquet Droz's Les Douze Villes combines a disk bearing the names of twelve cities with a jump-hour display (see the chapter on Retrogrades and Jump Hours). Simply pressing the pusher displays the required city and the corresponding time. Roger Dubuis's triple-time-zone Excalibur GlobeTimer model (which won Favorite Men's Watch 2006 from the Geneva Watchmaking Grand Prix) provides central main time and two 12-hour counters, each with 12 coordinating cities. Within the two counters, the larger hands indicate the hours, while the smaller hands indicate day or night. Fitted with a "city bezel" instead of an aperture, the Lange 1 Time Zone from A. Lange & Söhne is distinguished by its two offset sub-dials. Home time is generally displayed on the larger of the two, but an original adjustment system enables the user to reverse priorities by giving pride of place to local time (travel time) and by synchronizing it with the patented large date display.

### The world on the wrist

In 1936, Geneva-based watchmaker Louis Cottier invented an ingenious system enabling one to read off the time around the world at a glance. This principle has been developed and perfected in the famous Patek Philippe World Time models.

3

▸▸ 1 *Jaeger-LeCoultre, Master Hometime*
*The Master Hometime by Jaeger-LeCoultre is characterized by its crown-adjusted double hour-hand system.*

▸▸ 2 *Jaquet Droz, Les Douze Villes*
*On Les Douze Villes by Jaquet Droz, simply pressing the pusher enables one to display the desired city and its corresponding time.*

▸▸ 3 *A. Lange & Söhne, Lange 1 Fuseaux horaires*
*The Lange 1 Time Zones by A. Lange & Söhne is endowed with an original adjustment system enabling one to switch priorities between local (or travel) time and home time.*

*Vogard, Licensed Pilot*
*The Licensed Pilot by Vogard features an exclusive patented bezel-controlled time-zone adjustment system.*

1

▸▸ 1 *Patek Philippe, Heure Universelle Ref. 5130*
*In 2006, Patek Philippe presented a subtly redesigned version of its legendary Worldtimer watch.*

▸▸ 2 *Jaeger-LeCoultre, Master Geographic Worldtime*
*The Master Geographic Worldtime by Jaeger-LeCoultre provides the first-ever combination of a dual time zone with a world-time display.*

2

The new Worldtimer Ref. 5130, presented in 2006, is a worthy heir to this noble tradition. The central part of the dial features hands indicating local time. It is framed by two mobile disks, one graduated on a 24-hour scale (with day/night zones) and the other showing the names of 24 cities. To change the time zone, one need only press the single pusher at 10:00 and, by means of a patented system, the mechanism simultaneously adjusts the city disk, the 24-hour disk and the hour hand. In its Master World Geographic, Jaeger-LeCoultre combines a dual time-zone indication with a world-time display for the very first time. The crown at 2:00 enables one to shift the city disk to select the "priority" time zone that will appeal in a subdial at 7:00. In parallel, the user may read off the hour at any time in all 24 time zones. The Licensed Pilot by Vogard features an exclusive patented system designed to adjust the time zone by means of the bezel. The user simply selects the preferred city by turning the bezel, and the corresponding local time and day/night indication are automatically adjusted via an extremely sophisticated gear system. The bezel also enables one to take into account Daylight Saving Time or "summer" time when necessary.

## A. LANGE & SÖHNE

## LANGE 1 TIME ZONE

The Lange 1 Time Zone watch houses the Caliber L031.1 manually wound movement. Comprised of 417 parts and 54 jewels, this movement features twin mainspring barrels, a lever escapement and a balance spring with a frequency of 21,600 semi-oscillations per hour. The watch offers home time (hours, minutes, small seconds with stop seconds) with a day/night indicator, second time-zone indicator with day/night, and an outer city ring. Additionally there is a 72-hour power-reserve indicator, and a big-date indicator at 1:00. The easy-to-use watch features two off-center subdials. The larger subdial takes priority and displays home time. The smaller dial indicates time in another zone, which is set via a pushpiece. An ingenious adjustment mechanism makes it possible for the local time to become the home time when traveling, thereby reversing the priorities of the two subdials.

## AUDEMARS PIGUET

## JULES AUDEMARS DUAL TIME "ALL STARS" LIMITED EDITION

## REF. 26090PT.00.D028CR.01

Crafted in 950 platinum, this limited-edition timepiece features a second time zone displaying hours and minutes with AM/PM indication, power reserve and date. The dial is silvered and decorated with a star pattern and twelve Roman numerals. This 99-piece limited edition commemorates Arnold Schwarzenegger's After-School All Stars Foundation providing academic, recreational, cultural, and life skills for children in underprivileged environments.

## AUDEMARS PIGUET

## ROYAL OAK DUAL TIME - REF.26120OR.OO.D002CR.01

Crafted in 18K pink gold, this Royal Oak Dual Time houses the self-winding Caliber 2846. This timepiece features a second time zone displaying hours and minutes with AM/PM indication, power reserve and date. The dial is engine-turned with a Grande Tapisserie pattern and offers riveted gold indexes with facets. Also available in rose or yellow gold with silvered white dial.

## AUDEMARS PIGUET

## ROYAL OAK DUAL TIME - REF. 26120BA.OO.D088CR.01

This 18K yellow-gold Royal Oak Dual Time houses the self-winding Caliber 2846 and features a second time zone displaying hours and minutes with AM/PM indication, power reserve and date. The dial is engine-turned with a Grande Tapisserie pattern and offers riveted gold indexes with facets.

## AUDEMARS PIGUET

## ROYAL OAK DUAL TIME - REF. 26120ST.OO.1220ST.01

Crafted in stainless steel, this Royal Oak Dual Time houses the self-winding Caliber 2846 and features a second time zone displaying hours and minutes with AM/PM indication, power reserve and date. The dial is engine-turned with Grande Tapisserie pattern and offers riveted gold indexes with facets. Also available with black or navy dial.

## BLANCPAIN

## LE BRASSUS QUANTIÈME MOONPHASE GMT

This self-winding Frédéric Piguet caliber 67A6 for Blancpain features dual time with an independent center hand. From the Le Brassus collection, this model combines a full calendar (date, day of the week, month and moonphase) and the second time-zone display by an independent hand on a 24-hour basis. The movement consists of a self-winding 1151 caliber base and additional modules for full calendar and second time zone. The Quantième Moonphase GMT's case is water resistant up to 5atm.

## BLANCPAIN

## GMT ALARM TITANIUM

The Blancpain Calibre 1241 is now housed within a titanium case: The acoustics of this light and extremely resistant high-tech material will catch the ear of the most dedicated connoisseurs. The dial of the Léman Réveil GMT Titanium offers many useful indications: temporary residence time zone shown by the large central hour and minute hands; alarm hour and minute display at 3:00, 24-hour reference time at 9:00, small seconds at 6:00, date window between 4:00 and 5:00, alarm power reserve between 10:00 and 11:00, and finally alarm operating indicator between 12:00 and 1:00, activated by the crown at 8:00 via a column-wheel mechanism.

## BLANCPAIN

## LÉMAN 2100 TIME ZONE

The extra-thin, automatic self-winding Frédéric Piguet caliber 5L60 for Blancpain offers dual time on an additional dial with day/night indication and drives the Léman 2100 Time Zone model, whose water resistance is guaranteed up to 10atm. The dial's geometry is characterized by the off-center arrangement of the hour and minute display for the second time-zone, and by the presence of the relevant window for day/night indication. The crown controls all functions, including the date.

## BREGUET

## CLASSIQUE - REF. 5707BB/12/9V6

This Classique automatic wristwatch is crafted in 18K white or yellow gold, with self-winding alarm. Its engine-turned and silvered gold dial features a small seconds and date at 6:00, a second time zone on a 24-hour subsidiary dial at 9:00, an alarm time on a subsidiary dial at 3:00, an alarm power-reserve indicator between 10:00 and 12:00 and an alarm on/off indicator in a round aperture at 12:00. A crown at 2:00 winds the movement and the alarm and sets the time in two zones, another crown at 4:00 sets the alarm and the alarm on/off pushpiece is situated at 8:00. The water-resistant case is enhanced with a sapphire caseback.

## CHOPARD

## L.U.C TECH REGULATOR - REF. 16/8449

The L.U.C Tech Regulator has a mechanical manual-winding movement Chopard L.U.C 4RT. The pierced dial enables the technical details of the movement to be glimpsed. The regulator displays of minutes in the center, hours at 3:00 and seconds at 6:00 compose an astonishing geometry enlivened by a fine play of colors and by the use of Dauphine hands in hollowed-out blued steel. The display is completed at 9:00 by a second 24-hour time zone with day and night segments adjustable by means of a pushpiece. The L.U.C four-barrel movement, manufactured by Chopard in Fleurier, is COSC chronometer-certified and has a 9-day power reserve, which is displayed on the dial at 12:00. Its exclusive finish with fine straight line Côtes de Genève pattern and rhodium-plated décor can be admired through a transparent caseback. Limited and numbered series of 250 in steel.

## CHOPARD

## L.U.C PRO ONE GMT - REF. 16/8959

The L.U.C Pro One is equipped with an original GMT display system. Its distinctive feature lies in a bezel that turns at a click to indicate the hour in a second time zone on a 24-hour scale. Its steel case is water resistant to 300 meters. Bezel features orange SuperLumiNova numerals protected by a transparent sapphire crystal. This system provides a practical and original reading of the second time zone. The self-winding caliber, equipped with two stacked barrels (L.U.C Twin technology) and a power reserve of well over 65 hours, is chronometer certified by the COSC. The back is stamped with the four cardinal points in relief.

## DANIEL ROTH

## METROPOLITAN DUAL TIME - REF. 858.Y.50.172.CC.BD

This cosmopolitan timepiece is equipped with a self-winding exclusive and patented Daniel Roth mechanism composed of a GP 3100-based movement customized for Daniel Roth and enriched with an additional Daniel Roth plate. It is presented in a red-gold, white-gold or platinum case with open caseback, while the dial is adorned with guilloché Clous de Paris motif and flinqué center. It displays central hour, minute and second time-zone hands; 24 cities displayed at 12:00 and 6:00; central zone with AM/PM linked with second time zone, patented DST (summer/winter time) and day/night displays linked with reference time zone.

## DE GRISOGONO

## INSTRUMENTO DOPPIO TRE

This first double-faced watch with three time zones, the Instrumento Doppio Tre, combines a sophisticated movement with impeccable technical design. The watch offers two time-zone displays on one dial and one time on the other dial. All three displays work via a single-barrel movement comprised of more than 300 components. The case consists of two main parts, including a base cradle from which the case is lifted and rotated to reveal the second dial on the back. The front dial offers the double time-zone readout via superimposed counters and full date indication. The third time zone with analog hour and minute display is on the second dial, which is located uniquely under the rotor. The watch houses the mechanical ETA-based caliber with automatic-winding mechanism enhanced by de GRISOGONO.

## DE GRISOGONO

## INSTRUMENTO DOPPIO TRE

The reverse dial of the Instrument Doppio Tre reveals the third time-zone readout of this unique watch in an unusual manner: via hands that cross the oscillating width of the automatic movement because the dial is located under the rotor. The integrated mechanism for displaying a third time zone is a system patented by de GRISOGONO. The watch is water resistant to 50 meters.

## DE BETHUNE

## DB20 GMT

This watch is driven by a new self-winding caliber DB2024 with self-adjusting double barrel and innovative titanium/platinum rotor with optimal inertia-mass ratio, providing a 5-day power reserve. Its white-gold and palladium case measures 45mm in diameter features a titanium bezel with PVD silicon treatment. It also comprises an exclusive new system composed of a titanium balance bridge equipped with a triple shock-absorber with a double spring. The GMT mechanism shows the 24 hours in the dual time zone (jumping hours) linked to a day/night display. The light blued-steel dial with exclusive De Bethune hands enhances readability and the crocodile leather strap is fitted with a gold pin buckle or folding clasp.

## F.P. JOURNE

## CHRONOMÈTRE A RÉSONANCE

The Chronomèter Résonance is a stunning 40mm platinum-cased chronometer with an 18-karat gold dial and the FPJ caliber 1499-2 movement. The exclusive movement offers hours, minutes, small seconds and time in two zones, and 40-hour power reserve. It houses 267 components excluding the exterior mobile escapement holder. The watch consists of two barrels and two independent gear trains. The mechanism uses the natural resonance phenomenon to improve the watch's function.

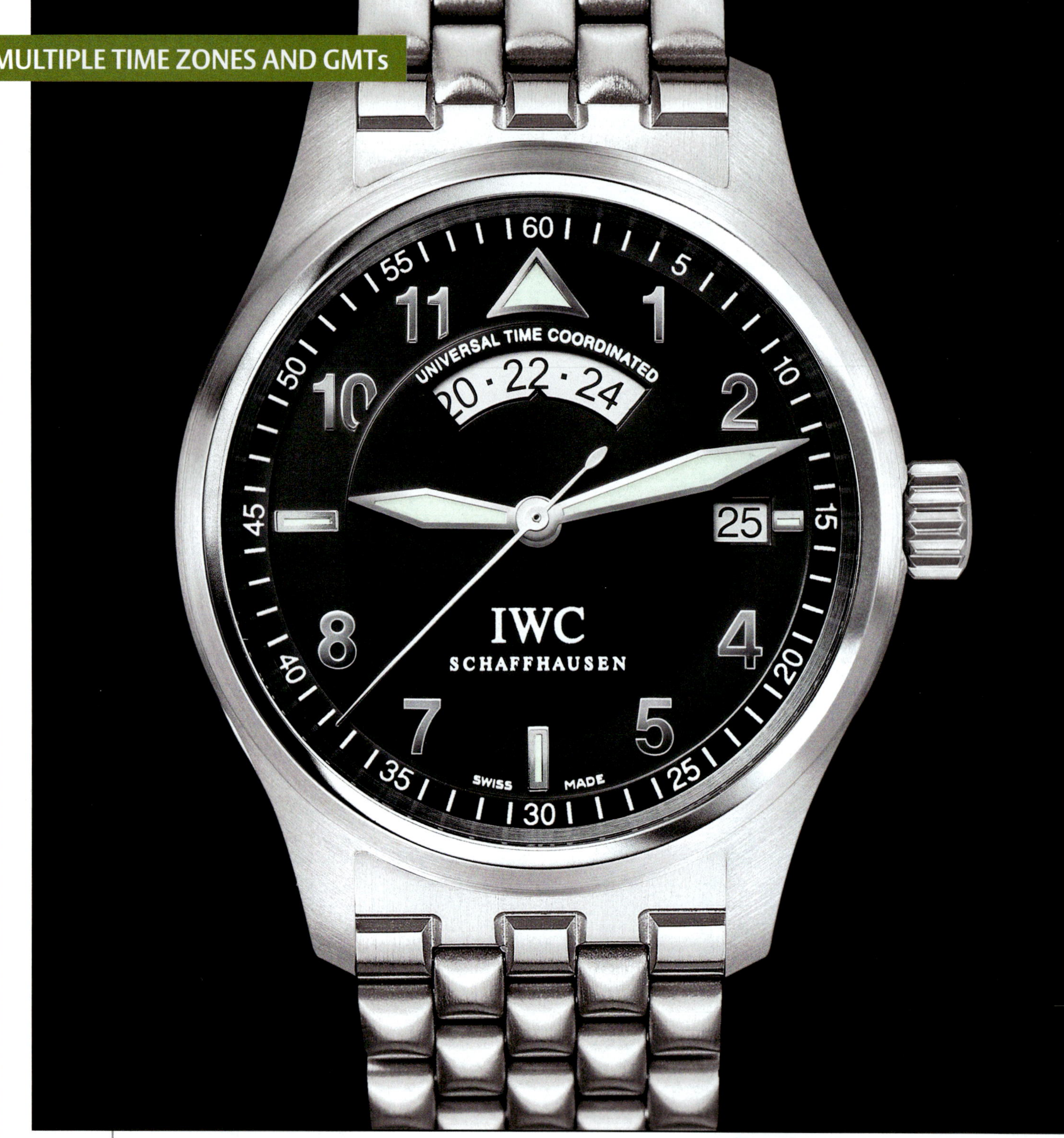


## IWC

## SPITFIRE UTC - REF. 3251

IWC's Spitfire UTC is crafted in steel and houses the caliber 37526 automatic movement, which oscillates at 28,800 vibrations per hour and features 21 rubies. It offers date indicator and 24-hour display. The time is adjusted in one-hour intervals and features a center second hand with stop device. The watch houses a soft iron inner case for protection against magnetic fields. It is water resistant to 6atm.

## JAEGER-LECOULTRE

## MASTER WORLD GEOGRAPHIC

This steel Master World Geographic is the latest in Jaeger-LeCoultre's line of multiple time-zone timepieces. On a single watch face is displayed the local time and all 24 time zones together with the precise time in one other selected time zone. This limited-series model is fitted with a new self-winding Jaeger-LeCoultre movement of remarkable toughness and reliability. The hours, minutes and seconds hands on the main dial track local time only, while a rotating dial ring and system of dial colors and shadings with colored arrows and sub-disks enable the easy legibility of universal and local timescales. An ingenious crown-operated mechanism makes it possible to highlight a major town in each time zone and it is even possible to distinguish between summer- and winter-time in both hemispheres. The date is displayed at 3:00 and there are indicators of the power reserve and months containing fewer than 31 days.

## JAEGER-LECOULTRE

## REVERSO SQUADRA HOMETIME

A worthy representative of the new Reverso sports watch line, the Squadra Hometime appears here in a steel case. It is powered by the automatic Jaeger-LeCoultre Caliber 977, which is equipped with a variable-inertia balance. It beats at a frequency of 28,800 vibrations per hour and boasts a 45-hour power reserve. To enhance shock resistance, the balance-bridge is fitted with two supports and the balance-spring is laser-welded to the stud and holder. The guilloché silver dial displays the hours, minutes, seconds and date, as well as a "smart" dual time zone with AM/PM indication. Water resistant to 50 meters and equipped with a sapphire crystal caseback, the Reverso Squadra Hometime is also available in 18K pink-gold and on a bracelet.

## JAEGER-LECOULTRE

## MASTER COMPRESSOR DUALMATIC

The Jaeger-LeCoultre Master Compressor Dualmatic is dedicated to the traveler. It offers local and 24-hour time. Elegantly housed in a 41.5mm case, the Dualmatic is fitted with the Caliber 972 high-performance automatic movement. It vibrates at 28,800 vibrations per hour and has 50 hours of power reserve. The movement is composed of 230 parts and 29 jewels. It offers hours, minutes, small seconds and date at the travel time, hours of reference time, 24-hour indicator synchronized with reference time and a rotating flange to protect against accidental rotation. It is fitted with one crown with compression key at 2:00 to adjust the rotating bezel and one crown with compression key at 4:00 for initial setting and starting, and moving the principal hour marker forward or backward. The caseback features an 18K gold 1000 Hours Control medallion.

## JAEGER-LECOULTRE

## MASTER HOMETIME

This Master Hometime watch houses Jaeger-LeCoultre's mechanical automatic movement 975 that beats at 28,000 vibrations an hour. It offers 50 hours of power reserve and features 230 parts and 29 jewels. The local hour hand (for travel time) can be moved forward or backward, date function is synchronized with the local hour hand, the home- or reference-time hour hand, minutes and small seconds hands, and day/night indicator. It is crafted in stainless steel or 18-karat pink gold and is water resistant to 5atm.

## JAEGER-LECOULTRE

## REVERSO DUOFACE NIGHT/DAY

The manual-winding JLC Caliber 854J offers dual time with a day/night indicator. A modification of Caliber 854 resulted in this edition of the Reverso Duoface whose unusual design of the main dial exhibits the day/night indication in a window placed at 12:00. Luminescent hands and enameled markers are combined, according to the brand's new aesthetic choices, with a black-enameled dial and stylized Chinese numerals in the Reverso models supplied with white-gold cases.

## JAEGER-LECOULTRE

## REVERSO GRANDE AUTOMATIQUE

For the first time, the Reverso Grande Automatique appears in a version fitted with an automatic movement, Jaeger-LeCoultre Caliber 970, a mechanical automatic movement with a barrel whose dimensions have been optimized to offer a 50-hour power reserve. The watch offers large date of travel time and second time zone with day and night indications. The 233-part movement vibrates at 28,800 beats per hour and is equipped with a unidirectional winding rotor mounted on ceramic balls requiring neither lubrication nor maintenance, a variable inertia balance to improve precision, and crown setting for all the indicators (local time, reference time, minutes, date, day/night display). It is offered in stainless steel or 18-karat pink gold.

## JAEGER-LECOULTRE

## REVERSO GRANDE GMT

The principle of two watches in one—such as the Reverso—has long delighted watch connoisseurs. The Reverso Grande GMT features a single movement driving back-to-back dials displaying different time zones. Crafted in stainless steel or 18-karat pink gold, the front dial displays hours, minutes, large date, small seconds, and day/night indicator. The dial has a silvered guilloché wave design. The contrasting black back dial features the 8-day power-reserve indicator and the symmetrically arranged day/night indicator, as well as the extraordinary GMT device. Reverso Grande GMT is powered by the manually wound Caliber 878 beating at 28,800 vibrations per hour and consisting of 276 parts and 35 jewels.

## JEAN-MAIRET & GILLMAN

## CONTINENTES

This square Continentes offers three retrograde hands and depicts time in three sections of the world: the Americas, Asia/Pacific and Europe/Africa. The mother-of-pearl dial of the watch has a world map decoration to indicate the three zones. Crafted in 18-karat rose gold, the watch houses the automatic JMG 3101TR/1153 caliber with 40 jewels, and is equipped with a 24-karat gold hand-decorated rotor.

## JEAN-MAIRET & GILLMAN

## HORA MUNDI

From the Around the World Collection, this Hora Mundi watch is crafted in stainless steel and houses the mechanical JMG 1999 self-winding movement. The watch offers 44 hours of power reserve, hours, minutes, seconds, date, alarm function and a world-time function via a disk. The Ø 41mm case is water resistant to 3atm. The black dial is accented with a silver cities zone and the watch is fitted with a sapphire caseback.

## MICHEL JORDI

## TWINS HERITAGE

Twins Heritage features two superimposed movements totalling 329 components, each in its own case, which slide together like a fan thanks to the ingenious patented "Twist-Lock" system. In the upper watch, the traditional, hand-winding MJ 1948.01 caliber with Geneva stripe finish and blued screws powers the functions of hours, minutes, small seconds, 42-hour power reserve, date and moonphase. An exclusive 3-layer dial displays moonphase and date and is available in ivory or ruthenium. In the lower watch, the 10" MJ 1948.02 caliber, the world's smallest traditional hand-winding, mono-pusher chronograph with column-wheel drive and also embellished with Geneva stripe finish and blued screws, powers the single pushbutton chronograph with 5/10" of a second indication, 30-minute counter and small seconds. An off-center chrono dial in white or anthracite shows a second time zone and is circled with a unique outer dial carved out of genuine Swiss alpine rock. Each 18K rose- or white-gold case is fitted with two antireflective, scratchproof, sapphire crystals and is water resistant to 30 meters.

## PATEK PHILIPPE

## WORLD TIME - REF. 5130

One of the easiest-to-use world time watches on the market, this World Time Ref. 5130 is crafted in 18K white gold and carries the Geneva Seal. The watch is powered by the mechanical automatic-winding Patek Philippe 240 HU caliber, which is decorated with a Côtes de Genève pattern and built with a 21K gold off-center micro-rotor. The watch offers hours, minutes, world-time, 24-hour and day/night indicators. The pusher at 10:00 enables adjustments of dual time by advancing the hour one hour at a time. The watch features a sapphire caseback and a screw-on caseback.

## PIAGET

## ALTIPLANO DOUBLE JEU - REF. G0A31152

This year, clothed in a new exterior and powered by new movements, the Altiplano Double Jeu watch reveals an unprecedented new face within this collection. Two superimposed pink-gold cases, each housing a mechanical hand-wound movement, are connected according to the principle of "secret" or case-spring watches. Simply pressing the pushpiece at 3:00 opens the upper case to reveal the second display. The upper case houses the Calibre 838P mechanical hand-wound movement driving the hours and minutes, as well as small seconds at 10:00; while the lower case beats to the cadence of Calibre 830P, a mechanical hand-wound movement indicating the hours and minutes.

## PIAGET

## EMPERADOR COUSSIN

Calibre 850P, a dual time-zone movement, has been incorporated within the Piaget Emperador Coussin watch, a model inspired by a vintage Piaget watch, available in pink gold or white gold and in a gem-set version. It is a mechanical self-winding movement measuring 12 lignes or 26.8mm in diameter and 4mm thick, endowed with a 72-hour power reserve. In addition to the large date window at 12:00, ensuring optimal readability, and a small seconds display, Piaget Emperador Coussin offers a function that is particularly useful these days: a second time zone, corresponding to travel time and appearing on a subdial at 7:00. Its center features a day/night indication synchronized with the central time zone, that of the place of residence.

## RICHARD MILLE

## RM 003-V2

The RM 003-V2 Tourbillon houses the mechanical, hand-winding movement 003-V2. In addition to the spline screws in grade-5 titanium for the bridges and case, many details exemplify Mille's technical approach to watchmaking such as the fast-rotating barrel with new involute tooth profile, central bridge in ARCAP, variable inertia PVD-treated balance wheel with overcoil balance spring and jewels set in white gold chatons. The finely finished pushbutton, located at 9:00, advances the second time zone in one-hour increments by utilizing a rotating disk of sapphire that allows the time zone to be visible at 3:00.

## RICHARD MILLE

## RM 015 PERINI NAVI CUP

The RM 015 Perini Navi Cup Tourbillon uses detailing inspired by nautical themes and the world of sailing. Many parts, such as the case design, crown and even the trademark Richard Mille titanium screws have undergone a transformation in this new RM model. It uses the manual-winding marine caliber 015, built on a carbon nanofiber baseplate and providing hours, minutes, indicators for power reserve and torque as well as a function indicator for Carica (winding), Neutrale (neutral) and Lanchette (hand setting). The finely finished pushbutton, located at 9:00, advances the second time zone in one-hour increments by utilizing a rotating disk of sapphire that allows the time zone to be visible near 10:30. Water resistant to 50 meters, the signature tripartite Richard Mille case with sapphire caseback is offered in titanium, 18K red or white gold and platinum.

## ULYSSE NARDIN

## MICHELANGELO GIGANTE UTC - REF. 22-11/41

This Michelangelo Gigante UTC, Ref. 223-11/41, is crafted in stainless steel and houses the UN-22 caliber. The watch offers a second time zone on the main dial with patented quickset, permanent home time at 9:00, and big date in a double window. The watch is also available in 18K rose gold, and with various dial combinations. The case measures 38x43mm, and the watch is available on a bracelet.

## ULYSSE NARDIN

## LADY DUAL TIME - REF. 223-22/30-09

The 37mm Lady Dual Time, Ref. 223-22/30-09, is crafted in stainless steel and houses the automatic winding UN-22 movement with 23 jewels and offering 42 hours of power reserve. The watch, with exhibition caseback, offers second time zone on the main dial with patented quickset, permanent home time in window at 9:00; big date in double window. It is available in 18K rose gold, with or without a diamond bezel, and can be purchased with a bracelet. It is water resistant to 100 meters

## ULYSSE NARDIN

## LADY DUAL TIME - REF. 223-28B/30-09

This Lady Dual Time, Ref. 223-28B/30-09, is crafted in a 37mm stainless steel case with an exhibition caseback. Housing the UN-22 caliber with automatic wind and 42 hours of power reserve, the watch offers second time zone on the main dial with patented quickset, permanent home time in window at 9:00, and big date in double window. It is also available in 18K rose gold, and with various dial combinations. The Lady Dual Time is available with or without a diamond bezel.

## ULYSSE NARDIN

## MICHELANGELO GIGANTE UTC - REF. 226-11/42

Crafted in 18K rose gold, this Michelangelo Gigante UTC, Ref. 226-11/42, houses the UN-22 automatic-winding movement in the 38x43mm case. The watch offers second time zone on the main dial, patented quickset, permanent home time in window at 9:00, and big date in double window at 2:00. It is also available in stainless steel, with various dial combinations, and on a bracelet.

## ULYSSE NARDIN

## DUAL TIME - REF. 243-55-7/92

This Dual Time, Ref. 243-55-7/92, is crafted in stainless steel, in a 42mm case. It houses the UN-24 movement and offers second time zone on main dial with patented quickset, permanent home time in window at 9:00, and big date in double window. With exhibition caseback, it is also available in 18K rose gold on either a strap or rose-gold bracelet.

## ULYSSE NARDIN

## QUADRATO DUAL TIME ZONE - REF. 243-92/601

This Quadrato Dual Time Zone, Ref. 243-92/601, is crafted in stainless steel in a 42x42mm case with exhibition caseback. The elegant piece offers second time zone on main dial with patented quickset, permanent home time in window at 9:00, and big date in double window. It is also available in 18K rose gold and with various dial combinations.

## ULYSSE NARDIN

## DUAL TIME - REF. 246-55/31

Crafted in 18K rose gold, this Dual Time, Ref. 246-55/31 is a 42mm-cased mechanical masterpiece powered by the UN-24. With exhibition caseback, the watch offers second time zone on main dial with patented quickset, permanent home time in window at 9:00 and big date in double window. It is also available in stainless steel, and each version comes with a bracelet option.

## ULYSSE NARDIN

## QUADRATO DUAL TIME - REF. 246-92/692

This Quadrato Dual Time, Ref. 246-92/692, is crafted in 18K rose gold and houses the UN-32 movement in its 42x42mm case. With exhibition caseback, it offers second time zone on main dial with patented quickset, permanent home time in window at 9:00, and big date in double window. It is available in stainless steel and with various dial combinations

## ULYSSE NARDIN

## SONATA CATHEDRAL DUAL TIME - REF. 670-88/213

Crafted in an 18K white-gold 42mm case, this Sonata Cathedra Dual Time, Ref. 670-88/213, houses the UN-67 movement with 109 jewels. The watch is equipped with a chiming 24-hour alarm with countdown indicator. It also features a dual time-zone system with instant time-zone adjustor. The watch offers 42 hours of power reserve and is water resistant to 30 meters.

## ULYSSE NARDIN

## SONATA CATHEDRAL DUAL TIME - REF. 676-88

Crafted in 18K rose gold, this Sonata Cathedral Dual Time, Ref. 676-88, houses the UN-67 caliber with automatic wind. The watch offers chiming 24-hour alarm with countdown indicator at 10:30, and a dual time-zone system with instant time-zone adjuster. The watch is available with various dial combinations and on a bracelet.

## VACHERON CONSTANTIN

## OVERSEAS DUAL TIME - REF. 47450/B01A

Encased in 316L surgical-grade steel on a steel bracelet, this Overseas Dual Time watch is powered by the mechanical automatic-winding Vacheron Constantin Caliber 1122 with 40 hours of power reserve. The 34-jeweled movement beats at 28,800 vibrations per hour, and is protected against magnetic fields by a soft inner iron case. The watch offers hours, minutes, seconds, date, and dual-time readout with day/night indication. Eight screws fasten the bezel and the sapphire crystal is antireflective. Water resistant to 15atm, Overseas Dual Time features a silvered dial with applied, luminescent gold trapezoid markers and luminescent hands.

## VACHERON CONSTANTIN

## MALTE DUAL TIME REGULATOR - REF. 42005

This Malte Dual Time Regulator is crafted in 18K pink gold in a half-hunter case with back lid, and houses the COSC-certified chronometer Vacheron Constantin Caliber 1206 RDT. The manually wound movement offers hours, minutes, date calendar at 6:00 and second time-zone indicator at 9:00 via a 24-hour subdial.

## VACHERON CONSTANTIN
## MALTE DUAL TIME - REF. 47400

This self-winding Malte Dual Time houses the Caliber 1222 and is available in 18K white or rose gold. It offers hours, minutes, date, power reserve, and second time-zone indication with day/night readout. The movement beats at 28,800 vibrations per hour; the watch is water resistant to 3atm.

## ZENITH

## GRANDE CLASS RESERVE DE MARCHE DUAL TIME ELITE

Housing the Elite 683 automatic extra-flat movement, this 44mm Grande Class Reserve de Marche Dual Time Elite consists of 192 parts and 36 jewels. Beating at 28,800 vibrations per hour, the watch offers 50 hours of power reserve (at 2:00) and features a 22K gold oscillating weight. In addition to hours and minutes, the watch gives small seconds, date and a 24-hour second-time-zone indication. Shown in 18K rose gold, it is also available in white and yellow gold, as well as steel.

# Moonphases

Beautifully depicting a small portion of the sky lighting up a dial, the moonphase indication is the most poetic and romantic of all complications. It serves as a glowing reminder of the timeless bonds between astronomy and the art of time measurement. Moonphase displays are often associated with other complications and appear on the dials of most perpetual calendar watches. Recently, we have even witnessed the emergence of several models giving pride of place to the night star, particularly on watches created exclusively for women.

### Celestial mathematics

With its nocturnal radiance and its ever-changing faces, the moon has always fascinated humankind. Most civilizations have ascribed to it magical influences on the rhythms of life and of nature. Its regular cycle inspired the monthly division of the year. In Western Europe, moonphase indications were an essential part of the first large medieval clocks. They are also featured on the first pocket watches, alongside other astronomical and astrological displays. In the 20th century, although the moon had lost much of its importance in planning daily life, watchmakers strove to miniaturize this mechanism to fit within a wristwatch while maintaining maximum precision—no small task indeed. The duration of a lunar cycle is 29 days, 12 hours, 44 minutes and 2.8 seconds (29.53 days). On ordinary moonphase indications, the disk bearing the two moons is driven by a 59-tooth wheel (two sets of 29.5).

1

2

This system results in a discrepancy of one day in 2 years, 7 months and approximately 20 days. High-end watches are fitted with a far more complex and accurate device called "astronomical moon," with a wheel comprising 135 teeth. On such models, the difference between the mechanism and the true lunar cycle amounts to only one day in 122 years! Such is the case for example with the L.U.C Lunar One by Chopard. This perpetual calendar featuring an original aesthetic appeal is distinguished by its "orbital" moonphase: the aperture through which the silver moon waxes and wanes actually spins about the axis of the small seconds, on a disk depicting a star-studded night sky. One year after releasing the Lange 1 Moonphase model, A. Lange & Söhne launched the Grande Lange 1 Lunar Mundi in 2003, offered as a pair of watches: a white-gold watch with the Big Dipper on the dial for the moonphase in the Northern hemisphere, and a pink-gold watch with the Southern Cross for the Southern hemisphere. The first "geopolitically correct" moonphase watch! Another key characteristic is that, instead of advancing once or twice every 24 hours, the moonphases are coupled to the hour wheel and progress in a regular manner. Meanwhile, the Portuguese Perpetual Calendar from IWC displays the moonphase in both hemispheres simultaneously, representing a clever way of using the two moons drawn on the disk. But the moonphase indication is first and foremost distinguished by its peerless precision: thanks to the large size of the self-winding movement, watchmakers were able to increase the number of teeth on the disk and thereby reduce the gap to an amazing one day in 577 years.

### New moons

Moonphases are generally displayed by means of a disk system appearing through a dial aperture. But some watchmakers prefer to give their imaginations free rein in order to invent new

ways of presenting the lunar cycle. De Bethune has equipped its Digitale watch with a patented spherical moonphase indication placed against a blue background occupying the entire back of the case. On its Les Lunes watch with complete day/date/month calendar, Jaquet Droz has opted for a pointer-type display, which is also equipped with a retrograde system: the hand follows the progression of the lunar cycle from one globe to the other (new moon, first quarter, full moon, last quarter, new moon), before jumping back to its point of departure. In its Equation Marchante watch (see the Equation of Time chapter), Blancpain also uses a retrograde pointer-type display. For its Kamar watch, Andersen Genève reinvented the traditional display mode with a round aperture above a mother-of-pearl disk comprising a dark moon and a light moon, depicted against a guilloché blue-gold dial, providing a particularly majestic larger-than-life display.

### Lady Moon

Due to its inherently poetic nature, the moonphase is considered the most feminine complication of them all. It is increasingly found on refined models destined exclusively for ladies' wrists. Perrelet gives it preferential placement in its Grande Phase de Lune Centrale with self-winding movement

6

▸▸ 1 *Chopard, L.U.C Lunar One*
*The L.U.C Lunar One by Chopard combines a perpetual calendar with an orbital moonphase that accumulates a discrepancy of just one day in 122 years.*

▸▸ 2 *A. Lange & Söhne, Grande Lange 1 Luna Mundi*
*The Grande Lange 1 Luna Mundi by A. Lange & Söhne comes in two versions: a white-gold watch for the moonphase in the Northern hemisphere, and a pink-gold watch displaying the sky of the Southern hemisphere.*

▸▸ 3 *IWC, Portugaise Perpetual Calendar*
*In the Portuguese Perpetual Calendar by IWC, the moonphase display deviates from astronomical reality by only one day in 577 years.*

▸▸ 4 *De Bethune, Digitale*
*The caseback of the Digitale watch by De Bethune carries the brand's patent ed spherical moonphase indication.*

▸▸ 5 *Jaquet Droz, Les Lunes*
*Jaquet Droz's Les Lunes is fitted with a retrograde pointer display.*

▸▸ 6 *Blancpain, Equation marchante*
*The Equation Marchante by Blancpain with its retrograde moonphase display.*

1

2

3

4

and pointer-type calendar; the moon's cycle stands out against a white mother-of-pearl dial sprinkled with diamond stars. Glamour is also the keynote on the Starmoon by Antoine Preziuso, presented as the largest moonphase ever created: the black ceramic dial with luminescent numerals is framed by a case set with diamond and sapphire stars. Patek Philippe has also pleased ladies with its Ref. 4958 with moonphases and small seconds. The moon appears through an aperture featuring an unusual size for a ladies' watch; the Geneva-based manufactures has accomplished an impressive feat by integrating moonphases and small seconds with the smallest mechanical movement carrying the prestigious Geneva Seal. On its Toric Rétrograde Perpetual Calendar Luna Blu, Parmigiani Fleurier presents a cosmic vision on the wrist with the indication of the lunar cycles in both hemispheres: Ursa Minor, Cetus, Lyra and Corona Borealis.

1 *Andersen Genève, Kamar*
*Andersen Genève reinvents the disk-type display on its Kamar watch, equipped with a majestic round aperture.*

2 *Antoine Preziuso, Starmoon*
*Antoine Preziuso presents its ladies' Starmoon, the largest moonphase display ever created.*

3 *Perrelet, Grande Phase de Lune centrale*
*Available in several versions, Perrelet gives pride of place to the night star on its Grande Phase de Lune Centrale for women.*

4 *Patek Philippe, Réf. 4958*
*The Patek Philippe Ref. 4958 ladies' watch combines a moonphase display with small seconds, driven by an extremely small movement.*

5 *Blancpain, Léman Réveil GMT Alarm*
*The alarm hand keeps step with the dual time-zone display, representing an extremely practical innovation on this Blancpain Léman GMT Alarm watch.*

5

# Alarm Watches

The ultimate useful function, an alarm may serve to help you wake up early, to remind you of an appointment, or to move your parked car before it gets towed away. In recent years, mechanical alarm watches have been enjoying renewed success, and this complication is also increasingly found within watches with multiple time zones—to the delight of businessmen and travelers who need to wake up on the other side of the world.

## BOVET

### 42MM ROSE-GOLD TRIPLE-DATE WITH MOONPHASE

This 18-karat rose-gold, self-winding, triple-calendar movement offers moonphase and sweep seconds. The watch features three disks under the dial that display the day, month and phase of the moon. Each indication has a corresponding correcting pushpiece. The movement is the self-winding caliber 11BA11, manufactured by Frederic Piguet according to Bovet's specifications. It consists of 28 rubies and beats at 28,800 vibrations per hour. The twin-barrel watch has 72 hours of power reserve, the 22-karat gold winding rotor is engraved in Fleurisanne style, while the bridges are beveled and decorated with the Côtes de Genève pattern. The watch is water resistant to 30 meters.

## JAEGER-LECOULTRE

## REVERSO GRANDE SUN MOON

Les généreuses proportions de ce garde-temps qui complète la collection Grande Reverso abritent le double barillet de son mouvement à remontage manuel Jaeger-LeCoultre de 873 qui garantit une réserve de marche de huit jours. Un pivotement du légendaire boîtier réversible ouvre la vue sur le mouvement artistiquement décoré à la main. Caractérisé par son centre finement guilloché, le cadran possède les chiffres stylisés emblématiques et les aiguilles en acier qui font partie du patrimoine génétique de la Reverso. Situé à 2 heures, l'indicateur jour/nuit fait disparaître le soleil à la fin de la journée afin de laisser monter dans le ciel une lune dont la phase est rappelée dans un petit cadran rond à 5 heures, partagé par l'aiguille des secondes. L'indicateur de réserve de marche se déploie comme un arc à 11 heures. Ce modèle est disponible en acier ou en or rose 18 carats sur un bracelet alligator avec boucle déployante.

## GIRARD-PERREGAUX

## SPORT CLASSIQUE LAUREATO EVO3 BIG DATE MOONPHASE - REF. 80185

Powered by the mechanical automatic-winding GP 3330-RLM00 caliber (GP 3300 caliber base modified for big-sized date, power reserve and moonphase module), this Sport Classique Laureato Evo3 Big Date Moonphase watch is crafted in steel and water resistant to 10atm. Beating at 28,800 vibrations per hour, the 35-jeweled movement has 46-hour power reserve and is decorated with the Côtes de Genève pattern. Functions including hour, minute, small second, big-sized date, and moonphase are indicated via luminescent hands, a red enamel second hand and a tachymeter scale.

## PIAGET

## EMPERADOR MOONPHASE TOURBILLON XL

This Tourbillon maintains the sophisticated lines and finishings of the Emperador watches, but offers a more powerful and masculine size. It houses the Piaget mechanical hand-wound Calibre 640P, the world's thinnest tourbillon movement at just 3.5mm thick. Admirably reflecting the mastery inherent to the manufacture, the carriage of this tourbillon—visible through a dial opening at 9:00—consists of 42 parts and weighs a mere 0.2 grams. A moonphase display appears at 6:00 on the rounded base of a delicately guilloché-worked surface. The lapis lazuli moonphase disk is enhanced by a golden moon and a pyrite star-studded dial.

# Glossary

ACCURACY s. Precision

ALARM WATCH (Above)
A watch provided with a movement capable of releasing an acoustic sound at the time set. A second crown is dedicated to the winding, setting and release of the striking-work; an additional center hand indicates the time set. The section of the movement dedicated to the alarm device is made up by a series of wheels linked with the barrel, an escapement and a hammer (s.) striking a gong (s.) or bell (s.). Works much like a normal alarm clock.

AMPLITUDE
Maximum angle by which a balance or pendulum wings from its rest position.

ANALOG or ANALOGUE
A watch displaying time indications by means of hands.

ANTIMAGNETIC
Said of a watch whose movement is not influenced by electromagnetic fields that could cause two or more windings of the balance-spring to stick to each other, consequently accelerating the rate of the watch. This effect is obtained by adopting metal alloys (e.g. Nivarox) resisting magnetization.

ANTIREFLECTION, ANTIREFLECTIVE
Superficial glass treatment assuring the dispersion of reflected light. Better results are obtained if both sides are treated, but in order to avoid scratches on the upper layer, the treatment of the inner surface is preferred.

ARBOR
Bearing element of a gear (s.) or balance, whose ends—called pivots (s.)—run in jewel (s.) holes or brass bushings.

AUTOMATIC (image 1)
A watch whose mechanical movement (s.) is wound automatically. A rotor makes short oscillations due to the movements of the wrist. Through a series of gears, oscillations transmit motion to the barrel (s.), thus winding the mainspring progressively.

AUTOMATON
Figures, placed on the dial or case of watches, provided with parts of the body or other elements moving at the same time as the sonnerie (s.) strikes. The moving parts are linked, through an aperture on the dial or caseback, with the sonnerie hammers (s.) striking a gong.

BALANCE (image 2)
Oscillating device that, together with the balance spring (s.), makes up the movement's heart inasmuch as its oscillations determine the frequency of its functioning and precision.

BALANCE SPRING (image 2)
Component of the regulating organ (s.) that, together with the balance (s.), determines the movement's precision. The material used is mostly a steel alloy (e.g. Nivarox, s.), an extremely stable metal compound. In order to prevent the system's center of gravity from continuous shifts, hence differences in rate due to the watch's position, some modifications were adopted. These modifications included Breguet's overcoil (closing the terminal part of the spring partly on itself, so as to assure an almost perfect centering) and Philips curve (helping to eliminate the lateral pressure of the balance-staff pivots against their bearings). Today, thanks to the quality of materials, it is possible to assure an excellent precision of movement working even with a flat spring.

BARREL (image 3-5)
Component of the movement containing the mainspring (s.), whose toothed rim meshes with the pinion of the first gear of the train (s.). Due to the fact that the whole—made up of barrel and mainspring—transmits the motive force, it is also considered to be the very motor. Inside the barrel, the mainspring is wound around an arbor (s.) turned by the winding crown or, in the case of automatic movements, also by the gear powered by the rotor (s.).

BEARING
Part on which a pivot turns, in watches mostly a jewels (s.).

BEVELING (image 4)
Chamfering of edges of levers, bridges and other elements of a movement by 45, a treatment typically found in high-grade movements.

BEZEL
Top part of case (s.), sometimes holds the crystal. It may be integrated with the case middle (s.) or a separate element. It is snapped or screwed on to the middle.

BOTTOM PLATE s. Pillar-plate

BRACELET
A metal band attached to the case. It is called integral if there is no apparent discontinuity between case and bracelet and the profile of attachments is similar to the first link.

BRIDGE
Structural metal element of a movement (s.)—sometimes called cock or bar—supporting the wheel train (s.), balance (s.), escapement (s.) and barrel (s.). Each bridge is fastened to the plate (s.) by means of screws and locked in a specific position by pins. In high-quality movements the sight surface is finished with various types of decoration.

BREGUET HANDS
A particular type of hands in a traditional elegant shape.

BRUSHED, BRUSHING
Topical finishing giving metals a line finish, a clean and uniform look.

CABOCHON (image 6)
Any kind of precious stone, such as sapphire, ruby or emerald, uncut and only polished, generally of a half-spherical shape, mainly used as an ornament of the winding crown (s.) or certain elements of the case.

CALENDAR, ANNUAL An intermediate complication between a simple calendar and a perpetual calendar. This feature displays all the months with 30 or 31 days correctly, but needs a manual correction at the end of February. Generally, date, day of the week and month, or only day and month are displayed on the dial.

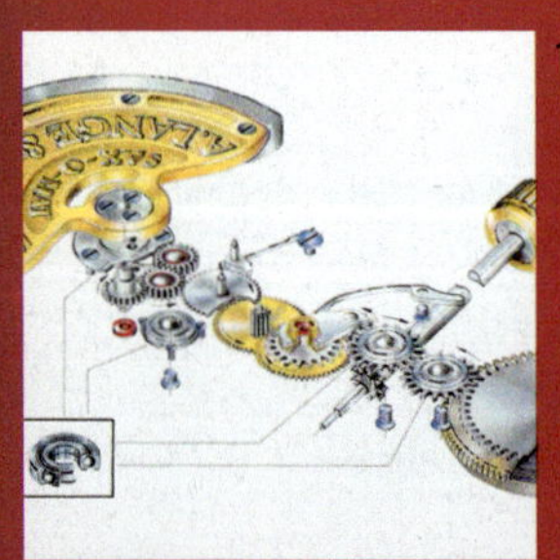

1

2

3

4

5

CALENDAR, GREGORIAN
With respect to the Julian Calendar (s. Calendar, Julian), the calendar reform introduced by Pope Gregory XIII in 1582 corrected the slight error of the former calendar by suppressing a leap year every hundred years, except for years whose numbers are divisible by 400 (this entailed the elimination of the leap years in 1700, 1800 and 1900, but not in 2000 and 2400). In non-Catholic countries this reform was introduced after 1700.

CALENDAR, FULL
Displaying date, day of the week and month on the dial, but needing a manual correction at the end of a month with less than 31 days. It is often combined with the moonphase (s).

CALENDAR, JULIAN
The calendar established by Julius Caesar was based on the year duration of 365.25 days with a leap year with 366 days every 4 years. In 325 AD, this calendar was adopted by the Church. Due to the slight error (0.0078 day) implied in this time count, the Julian Calendar was later replaced by the Gregorian Calendar (s. Calendar, Gregorian).

CALENDAR, PERPETUAL (image 7)
This is the most complex horology complication related to the calendar feature, as it indicates the date, day, month and leap year and does not need manual corrections until the year 2100 (when the leap year will be ignored).

CALIBER (image 8)
Originally it indicated only the size (in lines, ''') of a movement (s.), but now this indication defines a specific movement type and combines it with the constructor's name and identification number. Therefore the caliber identifies the movement.

CANNON
An element in the shape of a hollow cylinder, sometimes also called pipe or bush, for instance the pipe of the hour wheel bearing the hour hand.

CHAPTER-RING
Hour-circle, i.e. the hour numerals arranged on a dial.

CAROUSEL
Device similar to the tourbillon (s.), but with the carriage not driven by the fourth wheel, but by the third wheel.

CARRIAGE or TOURBILLON CARRIAGE (Top page)
Rotating frame of a tourbillon (s.) device, carrying the balance and escapement (s.). This structural element is essential for a perfect balance of the whole system and its stability, in spite of its reduced weight. As today's tourbillon carriages make a rotation per minute, errors of rate in the vertical position are eliminated. Because of the widespread use of transparent dials, carriages became elements of aesthetic attractiveness.

CASE (image 10)
Container housing and protecting the movement (s.), usually made up of three parts: middle, bezel, and back.

CENTER SECOND HAND, s. Sweep second hand.

CENTER-WHEEL
The minute wheel in a going-train.

CHAMPLEVÉ
Hand-made treatment of the dial or case surface. The pattern is obtained by hollowing a metal sheet with a graver and subsequently filling the hollows with enamel.

CHIME
Striking-work equipped with a set of bells that may be capable of playing a complete melody. A watch provided with such a feature is called chiming watch.

CHRONOGRAPH
A watch that includes a built-in stopwatch function, i.e. a timer that can be started and stopped to time an event. There are many variations of the chronograph.

CHRONOMETER
A high-precision watch. According to the Swiss law, a manufacture may put the word "chronometer" on a model only after each individual piece has passed a series of tests and obtained a running bulletin and a chronometer certificate by an acknowledged Swiss control authority, such as the COSC (s.).

CIRCULAR GRAINING
Superficial decoration applied to bridges, rotors and pillar-plates in the shape of numerous slightly superposed small grains, obtained by using a plain cutter and abrasives. Also called Pearlage or Pearling.

CLICK s. Pawl

CLOISONNÉ
A kind of enamel work— mainly used for the decoration of dials—in which the outlines of the drawing are formed by thin metal wires. The colored enamel fills the hollows formed in this way. After oven firing, the surface is smoothed until the gold threads appear again.

CLOUS DE PARIS
Decoration of metal parts characterized by numerous small pyramids.

COCK, s. Bridge.

COLIMAÇONNAGE (image 1 next page), s. Snailing.
COLUMN-WHEEL
Part of chronograph movements, governing the functions of various levers and parts of the chronograph operation, in the shape of a small-toothed steel cylinder. It is controlled by pushers through levers that hold and release it. It is a very precise and usually preferred type of chronograph operation.

COMPLICATION
Additional function with respect to the manual-winding basic movement for the display of hours, minutes and seconds. Today, certain features, such as automatic winding or date, are taken for granted, although they should be defined as complications. The main complications are moon-

6

7

8

9

10

phase (s.), power reserve (s.), GMT (s.), and full calendar (s.). Further functions are performed by the so-called great complications, such as split-second (s.) chronograph, perpetual calendar (s.), tourbilon (s.) device, and minute repeater (s.).

CORRECTOR
Pusher (s.) positioned on the case side that is normally actuated by a special tool for the quick setting of different indications, such as date, GMT (s.), full or perpetual calendar (s.).

COSC
Abbreviation of "Contrôle Officiel Suisse des Chronomètres," the most important Swiss institution responsible for the functioning and precision tests of movements of chronometers (s.). Tests are performed on each individual watch at different temperatures and in different positions before a functioning bulletin and a chronometer certificate are issued, for which a maximum gap of -4/+4 seconds per day is tolerated.

CÔTES CIRCULAIRES (image 1)
Decoration of rotors and bridges of movements, whose pattern consists of a series of concentric ribs.

CÔTES DE GENÈVE (image 2)
Decoration applied mainly to high-quality movements, appearing as a series of parallel ribs, realized by repeated cuts of a cutter leaving thin stripes.

COUNTER (image 5)
Additional hand on a chronograph (s.), indicating the time elapsed since the beginning of the measuring. On modern watches the second counter is placed at the center, while minute and hour counters have off-center hands in special zones (s.), also called subdials.

CROWN
Usually positioned on the case middle (s.) and allows winding, hand setting and often date or GMT hand setting. As it is linked to the movement through the winding stem (s.) passing through a hole in the case. For waterproofing purposes, simple gaskets are used in water-resistant watches, while diving watches adopt screwing systems (screw-down crowns).

CROWN-WHEEL
Wheel meshing with the winding pinion and with the ratchet wheel on the barrel-arbor.

DECK WATCH
A large-sized ship's chronometer.

DEVIATION
A progressive natural change of a watch's rate with respect to objective time. In case of a watch's faster rate, the deviation is defined positive, in the opposite case negative.

DIAL (image 6)
Face of a watch, on which time and further functions are displayed by markers (s.), hands (s.), discs or through windows (s.). Normally it is made of a brass—sometimes silver or gold.

DIGITAL
Said of watches whose indications are displayed mostly inside an aperture or window (s.) on the dial.

EBAUCHE (image 7)
Incomplete (jeweled or non-jeweled) watch movement without regulating organs, mainspring, dial and hands.

ENDSTONE (image 8)
Undrilled jewel, placed on the balance jewel with the tip of the balance-staff pivot resting against its flat surface, to reduce pivot friction. Sometimes used also for pallet staffs and escape wheels.

ENGINE-TURNED, s. Guilloché.

EQUINOX (image 9)
The time when day and night are of equal length, when the sun is on the plane of the equator. Such times occur twice in a year: the vernal equinox on March 21st-22nd and the autumnal equinox on September 22nd-23rd.

EQUATION OF TIME (image 4)
Indication of the difference, expressed in minutes, between conventional mean time and real solar time. This difference varies from -16 to +16 seconds between one day and the other.

ESCAPE WHEEL (image10)
A wheel belonging to the mechanism called escapement (s.).

ESCAPEMENT (image 5)
Positioned between the train (s.) and the balance wheel and governing the rotation speed of the wheel-train wheels. In today's horology the most widespread escapement type is the lever escapement. In the past, numerous types of escapements were realized, such as: verge, cylinder, pin-pallet, detent and duplex escapements. Recently, George Daniels developed a so-called "coaxial" escapement.

FLINQUÉ (Facing page, on the top)
Engraving on the dial or case of a watch, covered with an enamel layer.

FLUTED (image 1)
Said of surfaces worked with thin parallel grooves, mostly on dials or case bezels.

FLY-BACK (image 2)
Feature combined with chronograph (s.) functions, that allows a new measurement starting from zero (and interrupting a measuring already under way) by pressing down a single pusher, i.e. without stopping, zeroing and restarting the whole mechanism. Originally, this function was developed to meet the needs of air forces.

FOLD-OVER CLASP (image 3)
Hinged and jointed element, normally of the same material as the one used for the case. It allows easy fastening of the bracelet on the wrist. Often provided with a snap-in locking device, sometimes with an additional clip or push-piece.

FOURTH-WHEEL
The seconds wheel in a going-train.

1

2

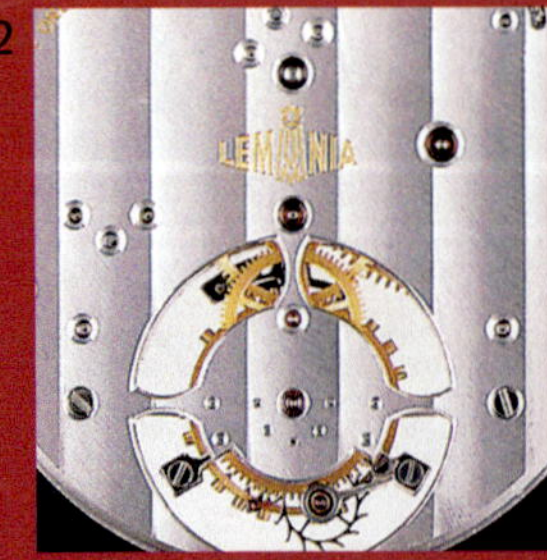

3

4

5

FREQUENCY, s. Vibration
Generally defined as the number of cycles per time unit; in horology it is the number of oscillations of a balance every two seconds or of its vibrations per second. For practical purposes, frequency is expressed in vibrations per hour (vph).

FUSEE
A conical part with a spiral groove on which a chain or cord attached to the barrel (s.) is wound. Its purpose is to equalize the driving power transmitted to the train.

GENEVA SEAL, s. Poinçon de Genève.

GLUCYDUR
Bronze and beryllium alloy used for high-quality balances (s.). This alloy assures high elasticity and hardness values; it is non-magnetic, rustproof and has a very reduced dilatation coefficient, which makes the balance very stable and assures high accuracy of the movement.

GMT
Abbreviation for Greenwich Mean Time. As a feature of watches, it means that two or more time zones are displayed. In this case, the second time may be read from a hand making a full rotation in a 24-hour ring (thereby also indicating whether it is a.m. or p.m. in that zone).

GOING TRAIN s. Train.

GONG
Harmonic flattened bell in a steel alloy, generally positioned along the circumference of the movement and struck by hammers (s.) to indicate time by sounds. Size and thickness determine the resulting note and tone. In watches provided with minute-repeaters (s.), there are often two gongs and the hammers strike one note to indicate hours, both notes together to indicate quarters and the other note for the remaining minutes. In more complex models, equipped also with en-passant sonnerie (s.) devices, there may be up to four gongs producing different notes and playing even simple melodies (such as the chime of London's Big Ben).

GRAND (or GREAT) COMPLICATIONS s. Complication

GUILLOCHé
Decoration of dials, rotors or case parts consisting of patterns made by hand or engine-turned. By the thin pattern of the resulting engravings—consisting of crossing or interlaced lines—it is possible to realize even complex drawings. Dials and rotors decorated in this way are generally in gold or in solid silver.

HAMMER
Steel or brass element used in movements provided with a repeater or alarm sonnerie (s.). It strikes a gong (s.) or bell (s).

HAND
Indicator for the analogue visualization of hours, minutes and seconds as well as other functions. Normally made of brass (rhodium-plated, gilded or treated otherwise), but also steel or gold. Hands are available in different shapes and take part in the aesthetic result of the whole watch.

HEART-PIECE
Heart-shaped cam generally used to realign the hands of chronograph counters.

HELIUM VALVE
Valve inserted in the case of some professional diving watches to discharge the helium contained in the air mixture inhaled by divers.

HEXALITE
An artificial glass made of a plastic resin.

HUNTER CALIBER
A caliber (s.) characterized by the seconds hand fitted on an axis perpendicular to the one of the winding-stem (s.).

IMPULSE
In a lever escapement (s.) the action of the escape-wheel tooth on the impulse face of the pallet; in the Swiss lever escapement it is produced by the impulse face of the wheel tooth and that of the pallet.

INCABLOC, s. Shockproof.

INDEX s. Regulator

JEWEL
Precious stone used in movements as a bearing surface. Generally speaking, the steel pivots (s.) of wheels in movements turn inside synthetic jewels (mostly rubies) lubricated with a drop of oil. The jewel's hardness reduces wear to a minimum even over long periods of time (50 to 100 years). The quality of watches is determined mainly by the shape and finishing of jewels rather than by their number (the most refined jewels have rounded holes and walls to greatly reduce the contact between pivot and stone).

JUMPING HOUR
Feature concerning the digital display of time in a window. The indication changes almost instantaneously at every hour.

LEAP-YEAR CYCLE
Leap or bissextile years have 366 days and occur every 4 years (with some exceptions, s. Calendar, Gregorian). Some watches display this datum.

LÉPINE CALIBER
A caliber (s.) typical for pocket-watches, characterized by the seconds hand fitted in the axis of the winding-stem (s.)

LIGNE s. Line.

LINE Ancient French measuring unit maintained in horology to indicate the diameter of a movement (s.). A line (expressed by the symbol ‴) equals 2.255mm. Lines are not divided into decimals; therefore, to indicate measures inferior to the unit, fractions are used (e.g. movements of 13‴3/4 or 10‴1/2).

LUBRICATION
To reduce friction caused by the running of wheels and other parts. There are points to be lubricated with specific low-density

6

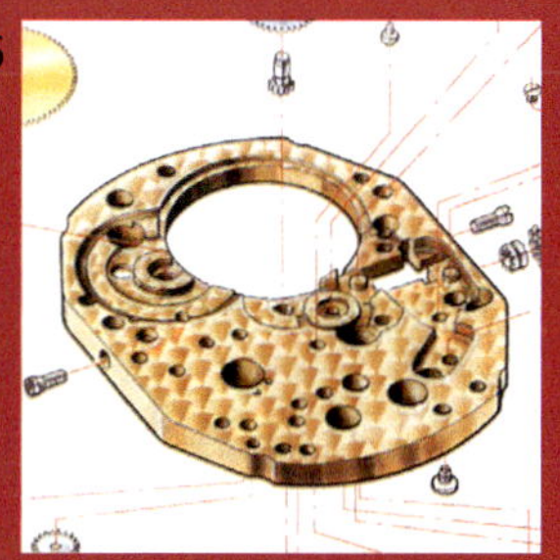
7

8

9

10

oils such as the pivots (s.) turning inside jewels (s.), the sliding areas between levers, and the spring inside the barrel (requiring a special grease), as well as numerous other parts of a movement.

LUG
Double extension of the case middle (s.) by which a strap or bracelet is attached. Normally, straps and bracelets are attached with removable spring bars.

LUMINESCENT
Said of materials applied on markers (s.) and/or hands (s.), emitting the luminous energy previously absorbed as electromagnetic light rays. Tritium is no longer used and was replaced by other substances having the same emitting powers, but with virtually zero radioactivity, such as Super-LumiNova and Lumibrite.

MAINSPRING
This and the barrel (s.) make up the driving element of a movement (s.). It stores and transmits the power force needed for its functioning.

MANUAL
A mechanical movement (v.) in which winding is performed by hand. The motion transmitted from the user's fingers to the crown is forwarded to the movement through the winding stem (s.), from this to the barrel (s.) through a series of gears (s.) and finally to the mainspring (s.).

MARINE CHRONOMETER
A large-sized watch enclosed in a box (therefore also called box chronometer) mounted on gimbals and used, on board of ships, to determine the respective longitude.

MARKERS
Elements printed or applied on the dial, sometimes they are luminescent (s.), used as reference points for the hands to indicate hours and fifteen- or five-minute intervals.

MEAN TIME
The mean time of the meridian of the Greenwich Observatory, considered the universal meridian, is used as a standard of the civil time system, counted from midnight to midnight.

MICROMETER SCREW (image 1)
Element positioned on the regulator, allowing to shift it by minimal and perfectly gauged ranges so as to obtain accurate regulations of the movement.

MICRO-ROTOR, s. Rotor. (image 2)

MINUTE REPEATER, s. Repeater.

MODULE (image 3)
Self-contained mechanism, independent of the basic caliber (s.), added to the movement (s.) to make an additional function available: chronograph (s.), power reserve (s.), GMT (s.), perpetual or full calendar (s).

MOONPHASE (image 3)
A function available in many watches, usually combined with calendar-related features. The moonphase disc advances one tooth every 24 hours. Normally, this wheel has 59 teeth and assures an almost perfect synchronization with the lunation period, i.e. 29.53 days (in fact, the disc shows the moonphases twice during a single revolution). However, the difference of 0.03 days, i.e. 44 minutes each month, implies the need for a manual adjustment every two and a half years to recover one day lost with respect to the real state of moonphase. In some rare case, the transmission ratio between the gears controlling the moonphase are calculated with extreme accuracy so as to require manual correction only once in 100 years.

MOVEMENT
The entire mechanism of a watch. Movements are divided into two great families: quartz and mechanical; the latter are available with manual (s.) or automatic (s.) winding devices.

MOVEMENT-BLANK s. Ebauche

NIVAROX
Trade name (from the producer's name) of a steel alloy, resisting magnetization, used for modern self-compensating balance springs (s.). The quality level of this material is indicated by the numeral following the name in decreasing value from 1 to 5.

OBSERVATORY CHRONOMETER
An observatory-tested precision watch that obtained the relevant rating certificate.

OPEN-FACE CALIBER s. Lépine Caliber.

OSCILLATION
Complete oscillation or rotation movement of the balance (s.), formed by two vibrations (s.).

OVERCOIL s. Balance spring.

PALLETS
Device of the escapement (s.) transmitting part of the motive force to the balance (s.), in order to maintain the amplitude of oscillations unchanged by freeing a tooth of the escape wheel at one time.

PAWL
Lever with a beak that engages in the teeth of a wheel under the action of a spring.

PILLAR-PLATE OR MAIN PLATE
Supporting element of bridges (s.) and other parts of a movement (s).

PINION
Combines with a wheel and an arbor (s.) to form a gear (s.). A pinion has less teeth than a wheel and transmits motive force to a wheel. Pinion teeth (normally 6 to 14) are highly polished to reduce friction to a minimum.

PIVOT
End of an arbor (s.) turning on a jewel (s.) support. As their shape and size can influence friction, the pivots of the balance-staff are particularly thin and, hence, fragile, so they are protected by a shockproof (s.) system.

PLATE s. Pillar-plate.

PLATED
Said of a metal treated by a galvanizing procedure in order to apply a slight layer of gold or another precious metal (silver, chromium, rhodium or palladium) on a brass or steel base.

PLEXIGLAS
A synthetic resin used for watch crystal.

POINÇON DE GENÈVE (image 4)
Distinction assigned by the Canton of Geneva to movements produced by watchmaker firms of the Region and complying with all the standards of high horology with respect to craftsmanship, small-scale production, working quality, accurate assembly and setting. The Geneva Seal is engraved on at least one bridge and shows the Canton's symbol, i.e. a two-field shield with an eagle and a key respectively in each field.

POWER RESERVE (image 5)
Duration (in hours) of the residual functioning autonomy of a movement after it has reached the winding peak. The duration value is displayed by an instantaneous indicator: analog (hand on a sector) or digital (through a window). The related mechanism is made up of a series of gears linking the winding barrel and hand. Recently, specific modules were introduced which may be combined with the most popular movements.

PRECISION
Accuracy rate of a watch, a term difficult to define exactly. Usually, a precision watch is a chronometer whose accuracy-standard is certified by an official watch-rating bureau, and a high-precision watch is a chronometer certified by an observatory.

PULSIMETER CHRONOGRAPH
The pulsimeter scale shows, at a glance, the number of pulse beats per minute. The observer releases the chronograph hand when starting to count the beats and stops at the 30th, the 20th or the 15th beat according to the basis of calibration indicated on the dial.

PUSHER, PUSH-PIECE or PUSH-BUTTON
Mechanical element mounted on a case (s.) for the control of specific functions. Generally, pushers are used in chronographs (s.), but also with other functions.

PVD
Abbreviation of Physical Vapor Deposition, a plating process consisting of the physical transfer of substance by bombardment of electrons.

RATCHET (WHEEL)
Toothed wheel prevented from moving by a click pressed down by a spring.

RATING CERTIFICATES s. Chronometer and COSC.
REGULATING UNIT (image 8)
Made up by balance (s.) and balance spring (s.), governing the division of time within the mechanical movement, assuring its regular running and accuracy. As the balance works like a pendulum, the balance spring's function consists of its elastic return and starting of a new oscillation. This combined action determines the frequency, i.e. the number of vibrations per hour, and affects the rotation speed of the different wheels. In fact the balance, by its oscillations, at every vibration (through the action of the pallets), frees a tooth of the escape wheel (s. Escapement). From this, motion is transmitted to the fourth wheel, which makes a revolution in one minute, to the third and then the center wheel, the latter making a full rotation in one hour. However, everything is determined by the correct time interval of the oscillations of the balance.

REGULATOR
Regulating the functioning of a movement by lengthening and shortening the active section of the balance spring (s.). It is positioned on the balance-bridge and encompasses the balance spring with its two pins near its fixing point on the bridge itself. By shifting the index, the pins also are moved and, by consequence, the portion of the balance spring capable of bringing the balance back is lengthened or shortened by its elastic force. The shorter it is, the more reactive it tends to be and the more rapidly it brings the balance back and makes the movement run faster. The contrary happens when the active portion of the balance spring is lengthened. Given today's high frequencies of functioning, even slight index shifts entail daily variations of minutes. Recently, even more refined index-regulation systems were adopted (from eccentric (s.) to micrometer screws (s.)) to limit error margins to very few seconds per day.

REMONTOIR, CONSTANT-FORCE
Old term used to denote any mechanism assuring a constant transmission of the driving power to the escape wheel.

REPEATER
Mechanism indicating time by acoustic sounds. Contrary to the watches provided with en-passant sonnerie (s.) devices, that strike the number of hours automatically, repeaters work on demand by actuating a slide (s.) or pusher (s.) positioned on the case side. Repeaters are normally provided with two hammers and two gongs: one gong for the minutes and one for the hours. The quarters are obtained by the almost simultaneous strike of both hammers. The mechanism of the striking work is among the most complex complications.

RETROGRADE
Said of a hand (s.) that, instead of making a revolution of 360 before starting a new measurement, moves on an arc scale (generally of 90 to 180) and at the end of its trip comes back instantaneously. Normally, retrograde hands are used to indicate date, day or month in perpetual calendars, but there are also cases of retrograde hours, minutes or seconds.

8

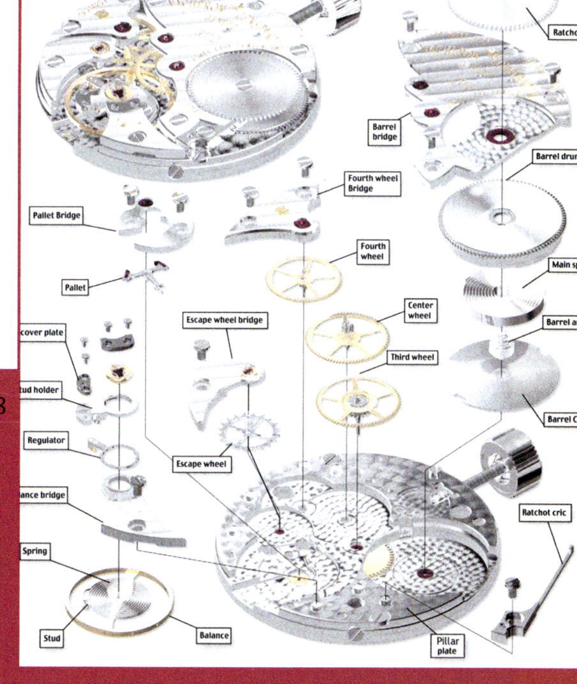

Unlike the case of the classical indication over 360, the retrograde system requires a special mechanism to be inserted into the basic movement.

ROLLER TABLE or ROLLER
Part of the escapement in the shape of a disc fitted to the balance staff and carrying the impulse pin that transmits the impulses given by the pallets to the balance.

ROTOR (image 1)
In automatic-winding mechanical movements the rotor is the part that, by its complete or partial revolutions and the movements of human arm, allows winding of the mainspring (s.).

SCALE (image 2)
Graduation on a measuring instrument, showing the divisions of a whole of values, especially on a dial, bezel. The scales mostly used in horology are related to the following measuring devices: tachometer (s.) (indicating the average speed), telemeter (s.) (indicating the distance of a simultaneously luminous and acoustic source, e.g. a cannon-shot or a thunder and related lightning), pulsometer (to calculate the total number of heartbeats per minute by counting only a certain quantity of them). For all of these scales, measuring starts at the beginning of the event concerned and stops at its end; the reading refers directly to the chronograph second hand, without requiring further calculations.

SECOND TIME-ZONE INDICATOR, s. GMT and World Time.

SECTOR, s. Rotor.

SELF-WINDING, s. Automatic.

SHOCKPROOF or SHOCK-RESISTANT (image 3)
Watches provided with shock-absorber systems (e.g. Incabloc) help prevent damage from shocks to the balance pivots. Thanks to a retaining spring system, it assures an elastic play of both jewels, thus absorbing the movements of the balance-staff pivots when the watch receives strong shocks. The return to the previous position is due to the return effect of the spring. If such a system is lacking, the shock forces exert an impact on the balance-staff pivots, often causing bending or even breakage.

SIDEREAL TIME
The conventional time standard refers to the sidereal year (defined in terms of an average of 365.25636 days) considered to be perfectly regular until very recently, but – even though this is not true – the difference is so slight that it is virtually neglected. As a unit of time, the sidereal day is used mainly by astronomers to define the interval between two upper transits of the vernal point in the plane of the meridian.

SKELETON, SKELETONIZED
Watches whose bridges and pillar-plates are cut out in a decorative manner, thus revealing all the parts of the movement.

SLIDE Part of a mechanism moving with friction on a slide-bar or guide.

SMALL SECOND
Time display in which the second hand is placed in a small subdial.

SNAILING (image 4)
Decoration with a spiral pattern, mainly used on the barrel wheel or on big-sized full wheels.

SOLAR TIME
Generally speaking, the time standard referred to the relative motion of the Earth and the Sun governing the length of day and night. The true solar day is the period measured after the Sun appears again in the same position from our point of observation. Due to the non-uniform rotation of the Earth around the Sun, this measure is not regular. As an invariable measure unit, the mean solar day corresponds to the average duration of all the days of the year.

SOLSTICE
The time when the sun is farthest from the equator, i.e. on June 21st (Summer solstice) and December 21st (Winter solstice).

SONNERIE (EN PASSANT)
Function consisting of an acoustic sound, obtained by a striking work made up of two hammers (s.) striking gongs (s.) at set hours, quarter- and half-hours. Some devices can emit a chime (with three or even four hammers and gongs). By a slide (s.) or an additional pusher (s.) it is possible to exclude the sonnerie device and to select a so-called grande sonnerie.

SPLIT-SECOND CHRONOGRAPH (image 5)
Chronographs with split-second mechanisms are particularly useful for timing simultaneous phenomena which begin at the same time, but end at different times, such as sporting events in which several competitors are taking part. In chronographs of this type, an additional hand is superimposed on the chronograph hand. Pressure on the pusher starts both hands, which remain superimposed as long as the split-second mechanism is not blocked. This is achieved when the split-second hand is stopped while the chronograph hand continues to move. After recording, the same pusher is pressed a second time, releasing the split-second hand, which instantly joins the still-moving chronograph hand, synchronizing with it, and is thus ready for another recording. Pressure on the return pusher brings the hands back to zero simultaneously, provided the split-second hand is not blocked. Pressure on the split pusher releases the split-second hand, which instantly joins the chronograph hand if the split-second hand happens to be blocked.

STAFF or STEM, s. Arbor.

STOPWORK
Traditional device (now obsolete) provided with a finger piece fixed to the barrel arbor and a small wheel in the shape of a Maltese cross mounted on the barrel cover, limiting the extent to which the barrel (s.) can be wound.

STRIKING WORK, s. Sonnerie and Repeater.

SUBDIAL, s. Zone.

SUPER-LUMINOVA, s. Luminescent.

SWEEP SECOND HAND A center second hand, i.e. a second hand mounted on the center of the main dial.

1

2

3

4

5

TACHOMETER or TACHYMETER (image 6)
Function measuring the speed at which the wearer runs over a given distance. The tachometer scale is calibrated to show the speed of a moving object, such as a vehicle, over a known distance. The standard length on which the calibration is based is always shown on the dial, e.g. 1,000, 200 or 100 meters, or—in some cases—one mile. As the moving vehicle, for instance, passes the starting-point of the measured course whose length corresponds to that used as the basis of calibration, the observer releases the chronograph hand and stops it as the vehicle passes the finishing point. The figure indicated by the hand on the tachometer scale represents the speed in kilometers or miles per hour.

TELEMETER (image 6)
By means of the telemeter scale, it is possible to measure the distance of a phenomenon that is both visible and audible. The chronograph hand is released at the instant the phenomenon is seen; it is stopped when the sound is heard, and its position on the scale shows, at a glance, the distance in kilometers or miles separating the phenomenon from the observer. Calibration is based upon the speed at which sound travels through the air, viz. approximately 340 meters or 1,115 feet per second. During a thunderstorm, the time that has elapsed between the flash of lightning and the sound of the thunder is registered on the chronograph scale.

THIRD WHEEL
Wheel positioned between the minutes and seconds wheels.

TIME ZONES
The 24 equal spherical lunes unto which the surface of the Earth is conventionally divided, each limited by two meridians. The distance between two adjacent zones is 15° or 1 hour. Each country adopts the time of its zone, except for countries with more than one zone. The universal standard time is that of the zero zone whose axis is the Greenwich meridian.

TONNEAU (image 7)
Particular shape of a watchcase, imitating the profile of a barrel, i.e. with straight, shorter, horizontal sides and curved, longer, vertical sides.

TOURBILLON (image 8)
Device invented in 1801 by A. L. Breguet. This function equalizes position errors due to changing positions of a watch and related effects of gravity. Balance, balance spring and escapement are housed inside a carriage (s.), also called a cage, rotating by one revolution per minute, thus compensating for all the possible errors over 360. Although this device is not absolutely necessary for accuracy purposes today, it is still appreciated as a complication of high-quality watches.

TRAIN (image 9)
All the wheels between barrel (s.) and escapement (s.).

TRANSMISSION WHEEL s. Crown-wheel

UNIVERSAL TIME
The mean solar time (s.) of the Greenwich meridian, counted from noon to noon, Often confused with the mean time (s.) notion.

VARIATION
In horology the term is usually referred to the variation of the daily rate, i.e. the difference between two daily rates specified by a time interval.

WINDING, AUTOMATIC s. Automatic

VIBRATION
Movement of a pendulum or other oscillating bodies, limited by two consecutive extreme positions. In an alternate (pendulum or balance) movement, a vibration is a half of an oscillation (s.). The number of hourly vibrations corresponds to the frequency of a watch movement, determined by the mass and diameter of a balance (s.) and the elastic force of the balance spring. The number of vibrations per hour (vph) determines the breaking up of time (the "steps" of a second hand). For instance, 18,000 vph equals a vibration duration of 1/5 second; in the same way 21,600 vph = 1/6 second; 28,800 vph = 1/8 second; 36,000 vph = 1/10 second. Until the 1950s, wristwatches worked mostly at a frequency of 18,000 vph; later, higher frequencies were adopted to produce a lower percentage of irregularities to the rate. Today, the most common frequency adopted is 28,800 vph, which assures a good precision standard and less lubrication problems than extremely high frequencies, such as 36,000 vph.

WATER RESISTANT or WATERPROOF (image 1)
A watch whose case (s.) is designed in such a way as to resist infiltration by water (3 atmospheres, corresponding to a conventional depth of 30 meters; 5 atmospheres, corresponding to a conventional depth of 50 meters.)

WHEEL
Circular element, mostly toothed, combines with an arbor (s.) and a pinion (s.) to make up a gear (s.). Wheels are normally made of brass, while arbors and pinions are made of steel. The wheels between barrel (s.) and escapement (s.) make up the so-called train (s.).

WINDING STEM
Element transmitting motion from the crown (s.) to the gears governing manual winding and setting.

WINDOW
Aperture in the dial, that allows reading the underlying indication, mainly the date, but also indications concerning a second zone's time or jumping hour (s.).

WORLD TIME (image 10)
Additional feature of watches provided with a GMT (s.) function, displaying the 24 time zones on the dial or bezel, each zone referenced by a city name, providing instantaneous reading of the time of any country.

ZODIAC
Circular belt with the ecliptic in the middle containing the twelve constellations through which the sun seems to pass in the course of a year.

ZONE
Small additional dial or indicator that may be positioned, or placed off-center on the main dial, used for the display of various functions (e.g. second counters).

# BRAND DIRECTORY

A. LANGE & SöHNE
Altenberger Strasse 15
D-01768 Glashütte
Germany
Tel: 49 (0) 35053 440
USA: 212 891 2355

AUDEMARS PIGUET
1348 Le Brassus
Switzerland
Tel: 41 21 845 14 00
USA: 212 758 8400

BLANCPAIN
Chemin de l'Etang 6
CH - 1094 Paudex
Switzerland
Tel: 41 21 796 36 36
USA: 201 271 1400

BOVET FLEURIER SA
9 rue Ami-Lévrier
1201 Geneva
Switzerland
Tel: 41 22 731 46 38
USA: 305 965 3277

BREGUET
1344 L'Abbaye
Switzerland
Tel: 41 21 841 90 90
USA: 866 458 7488

BVLGARI
34 rue de Monruz
2000 NeuchâTel
Switzerland
Tel: 41 32 722 78 78
USA: 212 315 9700

CHANEL
25 Place du Marché St Honoré
75001 Paris, France
Tel: 33 1 55 35 50 00
USA: 212 688 5055

CHOPARD
Rue de Veyrot 8
1217 Meyrin-Geneva 2
Switzerland
Tel: 41 22 719 31 31
USA: 212 821 0300

DANIEL ROTH
4, rue de la Gare
1347 Le Sentier
Switzerland
Tel: 41 21 845 15 55
USA: 212 315 9700

DE BETHUNE
6 Granges-Jaccard
1454 La Chaux L'Auberson Switzerland
Tel: 41 24 454 22 81
USA: 212 729 7152

de GRISOGONO
Route de St. Julien 176 bis.
1228 Plan-les-Ouates - Switzerland
Tel: 41 22 817 81 00
USA: 212 439 4240

EBEL
113 rue de la Paix
2301 La Chaux-de-Fonds - Switzerland
Tel: 41 32 912 31 23
USA: 201 267 8000

F. P. JOURNE
Rue l'Arquebuse 17
1204 Geneva, Switzerland
Tél: 41 22 322 09 09
USA: 1 305 531 2600

GÉRALD GENTA
4, rue de la Gare
1347 Le Sentier - Switzerland
Tel: 41 22 977 17 17
USA: 212 315 9700

GIRARD-PERREGAUX
Place Girardet 1
2301 La Chaux-de-Fonds
Switzerland
Tel: 41 32 911 33 33
USA: 201 804 1978

GREUBEL FORSEY
19-21 rue du Manège
2300 La Chaux-de-Fonds
Switzerland
Tel: 41 32 751 71 76
USA: 310 205 5555

GUY ELLIA
21 rue de la Paix
75002 Paris, France
Tel: 33 1 53 30 25 25

HARRY WINSTON
Rue de Lausanne 82
1202 Geneva - Switzerland
Tel: 41 22 716 29 00
USA: 1 212 245 2000

HAUTLENCE
6, Place des Halles
2000 NeuchâTel - Switzerland
Tel: 41 32 722 65 50
USA: 310 205 5555

HD3
Team Styling
Rue du Dr. Yersin, 10
1110 Morges - Switzerland
Tel: 41 21 802 51 88
USA: 310 205 5555

HUBLOT
44 Route de Divonne
CH 1260 Nyon - Switzerland
Tel: 41 22 990 90 00
USA: 800 536 0636

IWC
Baumgarten Strasse 15
8201 Schaffhausen - Switzerland
Tel: 41 52 635 65 90
USA: 212 891 2460

JACOB & CO
14 rue du Rhone, 4th Fl.
1204 Geneva, Switzerland
Tel: 44 22 819 18 42
USA: 212 719 5887

JAEGER-LECOULTRE
Rue de la Golisse 8
1347 Le Sentier Switzerland
Tel: 41 21 845 02 02
USA: 212 308 2525

JAQUET DROZ
Rue Jaquet Droz 5
2300 La Chaux-de-Fonds - Switzerland
Tel: 41 32 911 2888
USA: 201 271 4790

JEAN DUNAND
World Premiere Watchmaking SA
1228 Plan-les-Ouates, Geneva
Switzerland
Tel: 41 22 706 19 60
USA: 570 270 6160

JEAN-MAIRET & GILLMAN
11 Chemin du Petray
1222 Vesenaz, Switzerland
Tel: 41 22 703 40 20
USA: 561 651 7272

MAURICE LACROIX
Brandschenkestrasse 2
8039 Zurich, Switzerland
Tel: 41 44 209 11 11
USA: 800 794 7736

MB&F SA
Grange-Canal 35
1223 Cologny - Switzerland
Tel: 41 22 786 36 18

MICHEL JORDI
51, Chemin d'Eysins
1260 Nyon, Switzerland
Tel: 41 22 362 89 88 - USA: 310 205 5555

PARMIGIANI FLEURIER
Rue du Temple 11
2114 Fleurier, Switzerland
Tel: 41 32 862 66 30
USA: 949 489 2885

PATEK PHILIPPE
Chemin du Pont du Centenaire 141
1228 Plan-les-Ouates, Switzerland
Tel: 41 22 884 20 20
USA: 212 218 1240

PIAGET
37, Chemin du Champ-des-Filles
1228 Plan-les-Ouates - Switzerland
Tel: 41 22 884 48 44 - USA: 212 355 6444

PIERRE DEROCHE
Le Revers 1 - 1345 Le Lieu, Switzerland
Tel: 41 21 841 11 69

RICHARD MILLE
11 rue du Jura
2345 Les Breuleux Jura - Switzerland
Tel: 332 99 491900
USA: 310 205 5555

TAG HEUER
Louis-Joseph Chevrolet 6A
2300 La Chaux-de-Fonds - Switzerland
Tel: 41 32 919 80 00
USA: 973 467 1890

ULYSSE NARDIN
3, Rue du Jardin
2400 Le Locle - Switzerland
Tel: 41 32 930 74 00
USA: 561 988 8600

URWERK
34 rue des Noirettes
1227 Carouge-Geneva - Switzerland
Tel: 41 22 900 20 25
USA: 310 205 5555

VACHERON CONSTANTIN
Rue des Moulins 1
1204 Geneva, Switzerland
Tel: 41 22 316 17 40
USA: 212 713 0707

ZENITH
2400 Le Locle - Switzerland
Tel: 41 32 930 62 62
USA: 973 467 1890